STARTUP NCO

STARTUP NCO

Transform Your Enlisted Experience into a High-Growth Career

WILLIAM TRESEDER

STARTUP NCO
Transform Your Enlisted Experience into a High-Growth Career
First Edition

ISBN 979-8-9912039-5-1 *Hardcover*
979-8-9912039-4-4 *Paperback*
979-8-9912039-3-7 *Ebook*

To My Queen, who showed me what it takes to thrive in any situation. I love you. I love us.

CONTENTS

PART I. INTRODUCTION

PART II. BECOMING A STARTUP NCO

PART III. THE STARTUP CRUCIBLE

PART I

INTRODUCTION

My wife didn't know much about the military before we met. No one in her immediate family had worn the uniform. She went to an Ivy League college and then worked in finance and technology. Her experience was almost the exact opposite of mine. I had barely graduated from high school and had worked in fast food and construction before enlisting in the Marines. I deployed twice, once to Iraq and once to Afghanistan, while popping in and out of school. I was already in my early thirties and had only just finished college when my wife and I met; we were two people from completely different worlds.

She is a huge history buff, though. She and her dad used to watch World War II movies together, so she knew a lot about that era. She was struck by the heroism and moral imperative of the fight against the Nazis. This interest gave me the opportunity that I needed. I suggested we watch the HBO series *Band of Brothers* together and said it would help her understand more about me and my life in the Corps, especially since I was enlisted. The only people she knew in the military were officers.

I needed her to understand more about the pillars of my identity as a Marine noncommissioned officer (NCO), including the types of relationships I built and the hardships I faced. Those experiences shaped me in profound ways. She could only understand this part of me if I shared stories about military life, even if the stories weren't about me.

It was the right call. She loved every episode of the series. We spent a lot of time discussing the various people, their mindsets, and their behaviors. The group dynamics of the platoon gave me my first chance to explain the difference between officers and enlisted service members. She nodded along, but I could tell she didn't really understand what I was trying to say. It made sense to her that some people told others what to do, but she still saw everyone in the show as a bunch of guys in uniforms.

All these soldiers displayed incredible courage, creativity, and resilience. They were obviously brave, dedicated human beings who fought honorably in one of the most dangerous wars in history. They were molded by the incredible pressures of working together to accomplish nearly impossible missions. These men were experienced beyond their years, especially considering most were in their late teens or early twenties.

We binge-watched the series in a few days. The end credits included a biography of each soldier, focusing on what they did after the war. It had been a long time since I'd last watched *Band of Brothers*, and I had forgotten about this part. It ended up being a really helpful way to again try to explain what it means to be enlisted. I didn't need to remember the exact biographies of each soldier. I already knew what we were going to learn, but I still cringed a little bit as each person's life was summarized.

My wife was amazed and saddened to learn that none of the enlisted personnel ended up professionally successful. *Of course they didn't make it*, I thought. *They didn't really have a chance*. They were unable to adapt. Their military skills and mindset didn't automatically translate into lucrative, fulfilling careers. There was no readily available path for them to follow. They had opportunities for solid blue-collar work, but that did not offer thirty-plus years of increasing wages, responsibilities, and job satisfaction.

During the war, they had developed discipline, built rewarding relationships, and expanded their identity from local boys to war-fighting men. Yet in the peace that followed, the potential of these battle-tested young men was squandered. They all struggled in some meaningful way. Many couldn't hold down steady jobs in the postwar economy. Some destroyed relationships with the people they loved. A few died young, broken by battle and sucked into drugs and violence. The nation's first GI Bill took care of some returning service members, but many slipped through the cracks.

I read the same stories as my wife, but our reactions were different. I wasn't sad. In fact, I was afraid! Deep down, I worried that I would be one of those guys who showed promise in the military but never made anything of himself in civilian life. I didn't want to be the stereotypical enlisted tragedy. I wanted to be a success in every sense of the word.

ENLISTED PATHWAYS

In some ways, getting out of the military in 1945 was much different from the experience today. Back then, 15 million people, a full 10 percent of the population, were in uniform. Everyone knew at least one person who had served, either one of the millions from World War II or one of the millions from World War I. It was not hard for veterans to explain what they had done in the military. This was a blessing and a curse, as networks tended to form horizontally between officers and other officers, rather than vertically between officers and enlisted personnel.

There were different pathways available to transitioning service members at the end of World War II.[1] They could try for white-collar jobs, go blue collar, or work on a farm. For those who wanted to compete for one of the few

1 United States Census Bureau, "Historical Statistics of the United States, 1789–1945," accessed Sept. 24, 2024, https://www2.census.gov/library/publications/1949/compendia/hist_stats_1789-1945/hist_stats_1789-1945-chD.pdf.

white-collar jobs, college was affordable thanks to the GI Bill. Those who went blue collar still earned enough that they could raise a family. Almost half of them were unionized, which meant job security and long-term benefits. Agricultural employment had already peaked and was slowly declining, but some people still went to work on farms.[2]

What has happened since then? Several decades-long trends fundamentally reshaped America. Fewer than 6 percent of Americans are veterans, down from almost 20 percent in the early 1980s.[3] That is a decline of 70 percent, and it affects us as we navigate the labor market. In the past, one in five employees at a typical company was a veteran. Now, it's one in sixteen.

The lack of military experience in the civilian world can hurt your career unless you have a plan to address it. We have to figure out how to translate our skills and experience for an audience with zero understanding of the military. Unfortunately for us, the lazy way to explain the military is that the officers think strategically and

---

2 Christopher J. Tassava, "The American Economy During World War II," Economic History Association, accessed Sept. 24, 2024, https://eh.net/encyclopedia/the-american-economy-during-world-war-ii/.

3 Katherine Schaeffer, "The Changing Face of America's Veteran Population," Pew Research Center, Nov. 8, 2023, https://www.pewresearch.org/short-reads/2023/11/08/the-changing-face-of-americas-veteran-population/.

make decisions, while enlisted personnel carry out orders mindlessly. This view of us represents a significant hurdle to overcome—even if we are used to hearing "Thank you for your service."

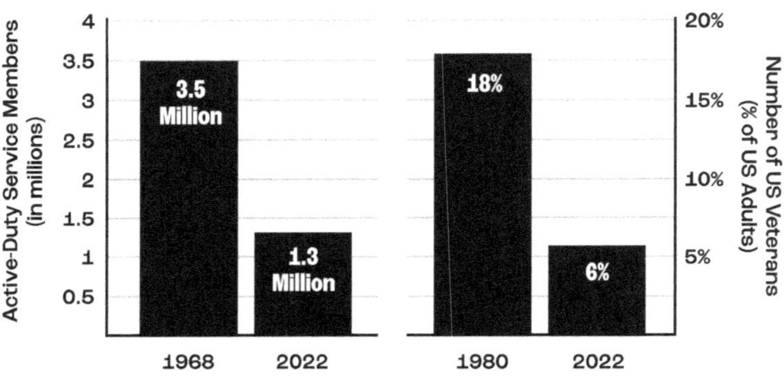

Source: Pew Research Center

At the same time, the job market has completely changed. Farm jobs have almost disappeared.[4] Manufacturing and construction peaked fifty years ago and have shrunk steadily since then, while jobs in the service sector—including health care, education, hospitality, and other businesses—have exploded.[5] Almost 15

4 Philip Martin, "Hired Farm Work Force Reports, 1945–87," *Farm Labor & Rural Migration News* (blog), Wilson Center, July 10, 2020, https://www.wilsoncenter.org/article/hired-farm-work-force-reports-1945-87.

5 Lyda Ghanbari and Michael D. McCall, "Current Employment Statistics Survey: 100 Years of Employment, Hours, and Earnings," *Monthly Labor Review*, US Bureau of Labor Statistics, August 2016, https://doi.org/10.21916/mlr.2016.38.

percent of jobs are now tied to federal, state, and local governments.[6]

Today, there is a much bigger divide between high-paying white-collar jobs and low-paying blue-collar jobs. In both cases, though, job security is a thing of the past. The only exception is government employment. The best bet is pursuing opportunities in local law enforcement, at the Department of Veterans Affairs, or at some other federal agency. Working in the government can scratch the itch of a stable job that includes some form of public service, but it's not likely to be a long-term solution to the problem of finding a challenging, lucrative career. These jobs tend to reward longevity rather than results. Those who want to push themselves usually leave. Those who remain have to fight against the tendency to get lazy and complacent.

There are lucrative white-collar jobs in industries such as finance and insurance. A few folks make those jobs work, but not many. I have enlisted buddies who made it through brutal Special Forces training and deployments. They were tough men, but they couldn't handle working at an investment bank. For many of us, it doesn't make sense to put in hundred-plus-hour weeks purely for the

---

6 Fiona Hill, "Public Service and the Federal Government," The Brookings Institution, May 27, 2020, https://www.brookings.edu/articles/public-service-and-the-federal-government/.

money. Even a temporary spike in salary isn't worth sacrificing our relationships with family and friends. There has to be something more to life than money, even if we want to make a lot of it.

Enlisted folks also continue to face barriers to leadership when we take off the uniform. Unlike officers, we are not given opportunities to step into leadership positions. We don't have access to the well-known pathways available to officers. Many corporate employers do not consider us a good fit for the jobs reserved for only the most educated, credentialed, and polished. We are rarely competitive when applying for top-tier colleges and universities that would set us up for long-term success.

There is a trap we have to avoid: we don't want to become low-level office workers playing around with spreadsheets or getting coffee for VIPs. Just because we can stomach long hours and poor working conditions does not mean that we should choose jobs with those characteristics. If you are happy with that sort of work, put down this book and go sign up for one of the "from the battlefield to the boardroom" programs.

Fortunately, there are other pathways for those of us who are willing to work smart and to work hard. We can still wake up and be on fire for a mission like we were in the military. We can keep building valuable skills. We can work in small teams with impressive folks. We can still make a positive impact for Americans and our allies. Life after the military can be as rewarding and fulfilling as life in it—maybe even more so.

STARTUP OPPORTUNITIES

Enter startups. This specific kind of high-growth, tech-powered business didn't exist after World War II. These companies emerged slowly, starting in Silicon Valley and Boston. Over the decades, people got rich building massive businesses such as Apple, Microsoft, Cisco, Google, Amazon, Facebook, Uber, and Netflix. The money and expertise created a startup ecosystem that churns out tight-knit teams working insanely hard on revolutionary products.

Now, here's the great news: as a hard-charging NCO willing to put in the work, you have all the raw materials you need to thrive at a startup. The military is a fantastic environment to develop many skills that are crucial for any dynamic, high-growth environment. NCOs are built to excel in chaos, trained to mold people into teams, and experienced and skeptical enough to cut through the unnecessary parts of complex plans. Fundamentally, NCOs know how to get the job done.

Hopefully you're reading this because deep down, you know you are special and you are willing to work very hard to build a great life. Maybe you're just starting to learn about startups but don't have firsthand experience. Maybe you've read some interesting articles about startups and want to see what all the fuss is about. Or maybe you already know you want to get a job at a startup but aren't sure how to break into that world as an NCO without a college degree.

Each type of person I just mentioned will learn a lot from this book. I've written it for NCOs who want to seize the chance to build great lives for themselves but aren't quite sure how. The single biggest factor in professional success for NCOs is breaking into the world of startups. Achieving that one thing will make all the difference, because it will place you in the best part of the economy. Startups are the massive businesses of the future. By growing quickly, they help employees build real wealth for themselves and their families.

GOALS FOR THIS BOOK

1. Get Job Offers
2. Adapt Military Skills

But you cannot break into startups simply by grinding away. This goal will require a strategic approach. Most

NCOs lack a process to do two very specific things: get fantastic job offers in high-growth startup companies and adapt the skills of a great NCO to that environment. Without a reliable process to identify and successfully pursue opportunities, you will miss out. Each great job will silently slip away while time passes and you fall farther behind.

THE STARTUP CRUCIBLE

The Startup Crucible laid out in this book is a roadmap to securing multiple job offers at incredible companies. It will help you build a diverse network of people who care about you and want to see you succeed as you transition to the private sector. Most importantly, this guide will upgrade your military mindset to a startup mindset. You will have the unique confidence that comes from experiencing a rapid transformation. You will build yourself into a Startup NCO.

Of course, there is plenty of work to do before that first startup job offer. You have to better understand yourself, including the ways that the military mindset and behaviors can hold you back in your professional career. You have to gain some knowledge about startups, including their history and specific ways of working. Startups are not military units, and you need to prepare yourself accordingly.

The first part of this book will help you think clearly about who you want to become, especially in the context of your military experience and goals for the future. That mindset shift is fundamental to your successful transition. Most of us have no idea what we are worth because we cannot see ourselves in comparison to others. Thanks to our experience as NCOs, we get better when the shit hits the fan. Those are the exact moments when most civilians tend to fall apart.

We also need to understand more about life as an employee. The roles I'm talking about are very different from the jobs you may have worked during high school for an hourly wage. Most of us don't know what it takes to succeed at business in the modern economy. As a result, we undervalue our NCO skills. The more we understand the practical, day-to-day challenges of businesses, the easier it is to imagine ourselves doing that work with the same confidence and energy that we had when leading teams in the military.

Once you commit to a positive vision of your future self, you can start to think about what you want to do when you grow up. Simply put, you need a job. And not just any job—you want to be selective about which job you choose. You have to resist the temptation to take the first acceptable one that comes along. The right first role upon leaving the military helps set the trajectory for your career. This means you need to have options. The second part of this book specifically covers the process that will get you at least five good job offers.

The job search period is critical, because it's when you'll bridge the gap between your NCO self and your new identity as a Startup NCO. This process will be a journey of self-discovery and rapid learning. You will drink from the fire hose as you pick up knowledge about different industries and job types, but that does not mean you will be overwhelmed. Following a step-by-step process with timelines and goals will help you stay on track, as will getting the help you need from family, friends, and new advocates you meet along the way.

The more you lean in to this transition, the better the next phase of your life will be. You will probably feel nervous as you start this first job out of the military. It will only take a few months, however, before you realize you are indispensable. The qualities you bring to the table are highly valued. Your coworkers and supervisors will learn to rely on you, to put their trust in you, and to give you greater responsibilities. That is the only real form of job security in the private sector.

All this may sound too good to be true, but it is absolutely possible for you. I know because I made this transition myself. It took me a long time and a lot of hard work, but I finally made it. I was fortunate to finish college in

Silicon Valley, where there were many opportunities to join startups. I stumbled through, learning many painful lessons along the way. Those hard-earned lessons are now yours, along with others that I picked up from other Startup NCO friends. Use this book to set big goals and then accelerate your transformation into the kind of person who can reach them.

MY JOURNEY

I joined the Marine Corps over twenty years ago. I have deployed to both Iraq and Afghanistan. And I have been in and out of the military three times. I'm currently back in the reserves for my fourth (and probably final) stint in the Corps. I have sat through the transition assistance workshops several times, used the GI Bill, and navigated the VA to receive a service disability, a home loan, and health care. Though I've found success, I've also made every mistake there is to make in the military and startups. I understand the challenges along the way, and I know what works.

Working my way through community college, I went from dropping out to earning a 4.0 GPA. After becoming the best student in my entire school, I transferred to Stanford University. From there, I maintained a high GPA while trying out a variety of jobs at startups, selecting from several job offers when I finished. I worked at a startup incubator (a place that builds new companies that then graduate to raise funding and grow into large businesses). I took a job at an early-stage startup, enjoying the ride before it crashed and burned less than a year later.

Then I decided to apply the lessons I'd learned to start my own company with several other veterans. I wanted to see if I had what it took to make it in the most intense business environment imaginable: a startup. I have since started four companies: one total failure called WorkScouts, one partial failure called VetCon, one partial success called NeuBridges, and one big success called BMNT. While building these companies, I've hired dozens of veterans (and decided not to hire many more hundreds who applied). I've helped veteran buddies launch their own startups. And I'm also an advisor to several venture capital funds that invest in startups.

During my years in the startup world, many folks have approached me about getting out of the military. Everyone from corporals to generals have asked me tough questions about what they should do as they take off the uniform. These initial questions have led to in-depth conversations about their goals, their dreams, their fears, and their identities. They've opened up to me, and I've walked them through effective ways to set goals and find jobs at startups. After hundreds of these conversations, I have learned the key topics to discuss. This book distills the essence of all that advice in one place.

So, what will these pages cover? I want to help you get a job at a high-growth startup while simultaneously adapting your military skills to help you crush that job from day one. I will tell you what you need to know, who you need to be, and what you need to do. The playbook is relatively simple to describe but hard to put into practice. You will have to let go of some parts of your military self that would hold you back in the next chapter of life. It's hard work—brutally hard work, actually, but 100 percent worth it.

This book is not meant to cover everything you could possibly learn on the subject. I will not talk about founding a startup. Go read *Zero to One* by Peter Thiel and *Shoe Dog* by Phil Knight. I will not talk about creating a small business. Go read *The E-Myth Revisited* by Michael Gerber. I will not talk about building a product. Go read *Build* by Tony Faddell and *Inspired* by Marty Cagan. I will not talk about getting jobs at huge technology companies such as Google or Amazon. And I will not talk about getting into—and graduating from—a great college. There are already lots of resources out there on all these topics, and you should take advantage of them.

This book is about breaking into the world of startups as an NCO, because I believe it's the best possible environment for high-potential people like us. The startup environment rapidly matures NCOs into leaders who will make a positive impact on the world. So, let's begin your journey to becoming a Startup NCO.

ONE

NCOs

The American colonies might have lost the battle for independence from Great Britain if not for the NCO. At the time, General George Washington was stuck with a ragtag bunch of state militias. They were called the Continental Army but did not function like one. They lacked any sense of military discipline and unit cohesion, each wearing different uniforms and using different weapons, tactics, and standards. Washington could not lead them as an effective fighting force, and he knew it. Something had to change; otherwise, the war would be over just as it was getting started.

General Washington turned to a Prussian military officer named Friedrich Wilhelm Ludolf Gerhard Augustin von Steuben (called Baron von Steuben for short). A veteran of the Seven Years' War and several other campaigns, von Steuben had proven in the past that he knew how to transform a group of individuals into a disciplined unit capable of winning battles. This was the hallmark of the Prussian military.

Washington knew talent when he saw it and decided to give von Steuben free rein to revamp his entire army. It helped that von Steuben was willing to serve without being paid, and he wasn't intimidated by the massive challenge of professionalizing militia men who were fighting the most powerful military in the world.

THE BIRTH OF THE NCO

Von Steuben's secret was the creation of an NCO Corps. Unlike in the British system, he ensured these men were selected based on ability. He introduced consistent standards that had nothing to do with social class. Soldiers were nominated to become NCOs based on their intelligence, work ethic, and ability to get results.

These NCOs were responsible for the daily grind that was so critical to the Continental Army's evolution. It included drilling troops, enforcing standards, engaging in physical training, and disciplining offenders. Von Steuben himself only trained the very first company of one hundred men. He expected this first group to serve as the model, representing the first "train the trainer" effort in the US military.[7]

Von Steuben created the modern distinction between officers and NCOs. The company commanders were ultimately accountable for their men but didn't train them. The specially selected NCOs were the ones responsible for getting the men up to common standards. To help these new NCOs as they drilled their own companies, von Steuben published *Regulations for the Order and Discipline of the Troops of the United States*. It became the standing field manual for over thirty years and set the standard for all manuals up to the present day.

The transformation between 1777 and 1780 was dramatic. Before von Steuben built this NCO Corps, all of General Washington's ideas were just that: ideas. The NCOs became the mechanism for enforcement. This dynamic planted the seed that has since grown into such a central part of the NCO's life: NCOs operationalized Washington's plans, turning them into reality on the battlefield. And now, hundreds of years later, we continue to do exactly the same thing.

---

7 Brian M. Shay, "After 230 Years, the 'Blue Book' Still Guides NCOs," US Army, Nov. 2, 2009, https://www.army.mil/article/29717/after_230_years_the_blue_book_still_guides_ncos.

All of us know the basics about how the Revolutionary War came to an end. Thanks to the newfound competence of the Continental Army, Washington was able to hold out for years against the superior British force. Eventually, the French came to our aid with the only other big navy in the world. Surrounded, the British were forced to surrender and withdraw, setting the stage for America's full independence. The culture, traditions, and freedoms we enjoy today can all be traced back to the endurance and courage of Washington's Continental Army.

THE MILITARY'S BACKBONE

From the Revolutionary War to World War II, NCOs continued to adapt and evolve. This was a period of extremes. The size of the military grew and shrank according to the nation's needs, but NCOs remained the backbone through it all. When the military grew, NCOs helped fresh recruits coalesce into a fighting force. They updated tactics after each battle in response to their experiences. And they kept the fighting spirit alive, no matter how bad things got.

NCOs were just as important in peacetime. Someone needed to maintain the institutional memory of the military and to preserve the identity and pride of the various units. This was a critical part of the NCO's role: they served as the middle layer, "translating" between the officers and enlisted personnel, while helping the organization turn ideas into reality. Without them, the military would have to start from scratch each time the nation was called to defend itself.

For over 150 years, NCOs maintained their role in the cycle of growing and shrinking. An endless series of wars caused spikes in the size of the military, followed by the inevitable shrinkage when peace returned. Wave after wave had the same pattern: The military grew to fight the War of 1812 and then shrank. It grew again during the Civil War, then shrank back down. The cycle repeated in the Spanish-American War, the Philippine Insurrection, the Banana Wars, and finally World War I.

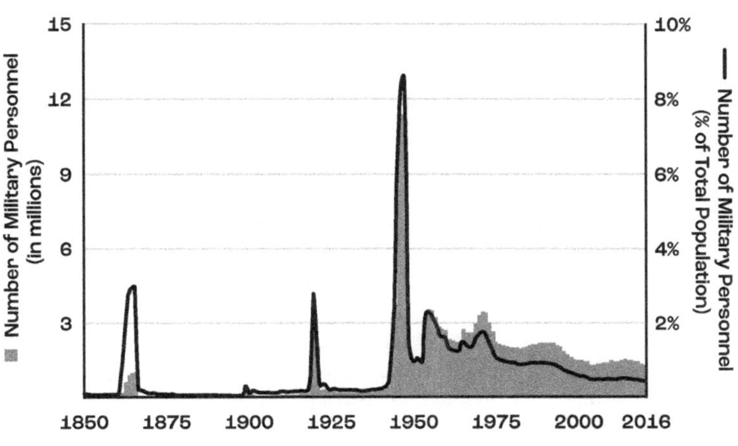

Source: Statista

Then something unexpected happened: following World War II, the US became a military superpower with global responsibilities. The military followed its usual pattern as the country prepared for war, growing almost fortyfold, from under 350,000 in 1939 to over 12 million in 1945.[8] It shrank to about 1.5 million before growing again for the Korean War. It then stabilized at around 3 million, which was ten times the size it had been less than ten years prior.[9] The United States went from a regional power with a strong economy to one of the world's two superpowers. All of a sudden, we had bases around the globe. That meant we needed a military large enough to man them, which led to the first peacetime draft in our history.

PROFESSIONALIZING AND SPECIALIZING

NCOs evolved to fit the new needs of the country. We became more technical, with support roles designed to

8 The National WWII Museum, "Research Starters: US Military by the Numbers," accessed Sept. 24, 2024, https://www.nationalww2museum.org/students-teachers/student-resources/research-starters/research-starters-us-military-numbers.

9 Wikipedia, "Demobilization of United States Armed Forces after World War II," last modified July 1, 2024, 18:36 (UTC), https://en.wikipedia.org/wiki/Demobilization_of_United_States_Armed_Forces_after_World_War_II#:~:text=The%20unpopular%20draft%20was%20terminated,million%20to%20about%201.5%20million.

take care of sophisticated weapons systems. We became more specialized, developing niche skills as new military occupations emerged. We became more managerial, implementing complex policies such as the racial integration plans of the Truman administration. We even became more political, learning to live among foreign populations and represent the United States.

Toward the end of the Vietnam War, the NCO's role evolved one more time. America decided to get rid of the draft and shift to an all-volunteer force. This was the last big change that brought us into the modern era. The military learned how to recruit and rapidly train people from America's high schools and colleges. If someone decided to reenlist, then they were on the NCO track. The military provided training and educational programs that put us on par with our civilian counterparts. In terms of leadership opportunities, we actually outpaced our civilian peers. We were put in charge of teams earlier in our careers, gained experience in dangerous training and combat environments, and dealt with a wider array of leadership challenges—including physical fitness, financial literacy, and moral education.

In short, NCOs became true professionals in the modern sense of the word. We aren't the short tour folks. Those are the junior enlisted. And we aren't officers, who come from pedigreed backgrounds and successful academic careers.

NCOs are something else entirely. We serve as a distinct layer in the military hierarchy, one that is critical to accomplishing any mission. We evolved from doers into translators between the thinkers and the doers. We take the commander's intent and turn it into reality. We manage up and manage down. We develop a keen sense of when people need to be pushed and when they need to be supported. And we know how to simplify to the point of clarity without sacrificing strategic value.

This is the heritage of a modern NCO in the American military. Membership in this community is priceless, and it is an incredible foundation that you can use to build a fulfilling, lucrative, impactful life after taking off the uniform.

UNDERSTAND YOUR HERITAGE

It takes a lot of time and thought to get out of the military. You have to discover a path from where you are today to success in the private sector. A big part of this transition is understanding your heritage as an NCO. You have successfully filled a specialized role that is critical for the military to adapt to dynamic environments.

That said, adapting to life after the military is hard, and adapting to life at a startup is even harder. NCO skills are only part of the equation. Fortunately, the adaptation is hard in the same way that life as an NCO is hard. We are talking about "running a marathon" hard, not "doing quantum physics in your head" hard.

Part II of this book lays out a proven plan to succeed on this hard path. But before we begin that work, we need to get a better sense of our destination. We need to understand the world of startups. The startups that you will work for are entirely different from the large, successful technology companies like Google or Uber that have already graduated to the big leagues. To understand the kinds of startups I'm talking about, rewind the clock twenty-five years for Google or fifteen years for Uber, and you will find two very different companies. In the early years, both organizations featured small groups of hard-working, slightly crazy people grinding away at a revolutionary idea. Every day was chaos, and no one knew whether the company would survive. This is the kind of place you want to be: where the future is being built and fortunes are being made.

TWO

STARTUPS

Love him or hate him, Christopher Columbus made his mark on history. He wanted to blaze a trail around the world, hoping it would make him rich and famous. Columbus had a vision but no money. He had two things going for him, though: being an experienced sailor and a damn good salesman. Using those two traits, he persuaded King Ferdinand and Queen Isabella, the monarchs of Spain, to fund his crazy idea of sailing west to get to the Indies. People were making huge fortunes there, and Columbus wanted to get in on the action.

Armed with this royal money, Columbus secured ships and recruited enough sailors to man them. The news of his successful first expedition triggered a wave of investment in more voyages, reshaping the entire western hemisphere. Thousands of ships sailed across the Atlantic over the next few hundred years. Some people made huge fortunes, while millions of others died from disease, starvation, abuse, war, and slavery.

What does a fifteenth-century explorer have to do with twenty-first-century technology companies? More than you might think. High-growth technology startups share a surprising number of economic similarities with the risky European expeditions that sailed six hundred years ago.

Let's begin by defining a few important terms to help clarify these parallels and get you more comfortable with the idea that you are actually well-suited to work at a startup. By reading this chapter and remembering these insights, you will already be ahead of most people trying to break into the world of startups.

RISK AND REWARD

Most people think of the words *technology* and *startup* as inextricably linked. You can't have one without the other, right? Wrong. If you want to thrive in a startup environment, correct this misconception right now.

Startups are businesses. Technology may be an important part of those businesses, but it is not the business itself. Every business exists to do one thing: make money. No money, no business—even if you have the coolest technology the world has ever seen. It's true that technology is often the critical factor in a business because it helps efficiently scale products and services. That means higher profit margins for the business and higher returns for the investors. The more money investors make, the more money they have to invest in companies. It's a virtuous cycle. Technology accelerates the cycle, but it doesn't magically make it work. A crappy business with a lot of cool technology is still a crappy business.

Startups are a specific kind of high-risk, high-reward business. Like other businesses, their goal is to make a lot of money, ideally for long periods of time. Unlike a lot of other businesses, they do not have a proven way to make all that money, which is how startup founders are like Christopher Columbus. They are trying to *explore*. They have a big vision that, if true, would make them and their investors rich. They need to figure out how to make their vision a reality. Like Columbus, these founders face a huge amount of risk and uncertainty along with the potential for a massive profit if they're right. Also like Columbus, they lack the resources required to do all the work on their own, and they will trigger waves of unpredictable consequences as they change the world.

STARTUPS ACROSS HISTORY

Drawing these historical parallels will help you understand the financial logic of the startup world. It is easier to understand using the simpler models from the past. After all, high-risk, high-reward opportunities have existed for a long time. The justification for many military expeditions has been the potential to take resources—human and otherwise—from the conquered people.

Columbus's basic idea was not a new one. We have evidence of such expeditions dating back thousands of years in every major civilization. Phoenician merchants were allowed to keep 20 percent of the cargo they carried from one place (where it was cheap) to another place (where it was expensive). Vikings had a system of distributing spoils across a raiding party when they returned from successful raids (meaning they killed people and stole their valuable stuff). During the Renaissance, Italians had a specific type of contract called a *commenda* or *colleganza*[10] that helped risky expeditions raise fund-

10 Frederic Chapin Lane, Venice and History: The Collected Papers of Frederic C. Lane (Baltimore: Johns Hopkins University Press, 2020), 59, https://dx.doi.org/10.1353/book.71823.

ing by sharing a percentage of the profit, assuming there was any.

But something important changed after Columbus. Sixteenth-century British and Dutch investors and entrepreneurs took the next logical step: they created the joint-stock company, which provided legal and financial protection against loss. Up to that point, investors could come after the personal assets of an entrepreneur if, for example, their ship was lost at sea. This made everyone more cautious, limiting trade and economic growth.

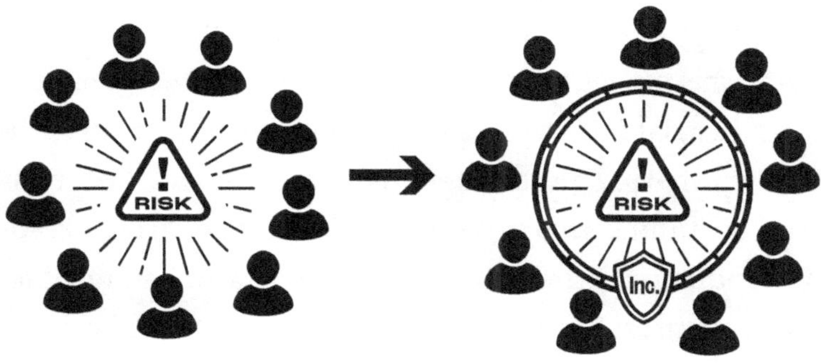

The invention of the joint-stock company addressed a serious issue of risk. This legal structure allowed multiple investors to buy shares—with an equivalent portion of the profits—in a high-risk, high-reward expedition. The approach spread the risk of loss across a greater pool of investors by making it easy to "buy in," while also limiting the liability of the entrepreneur who was putting together the deal. These new legal tools increased incentives to trade, driving greater economic growth.

The joint-stock, limited-liability strategy was quickly adopted all over Europe and soon spread to America. It offered the perfect vehicle to fund the expansion of new types of businesses.[11] The model was so successful that it was permanently made law in 1855–1856 by the United Kingdom's Limited Liability Act,[12] and then by America and other European nations soon after. New England whaling fleets used it to fund their incredibly profitable operations in the eighteenth and nineteenth centuries. During the Industrial Revolution, massive companies were built using this model, including US Steel and J.P. Morgan.

Industrial growth was not all positive. While this process of sharing profits and reducing risk worked very well to fund speculative investments and drive explosive industrial growth, it unfortunately also had terrible consequences. Industrialization contributed to atrocities such as environmental destruction, brutal child labor, and chattel slavery. Bad stuff does not get better as it gets bigger; it only gets worse.

---

11 Ron Harris, "A New Understanding of the History of Limited Liability: An Invitation for Theoretical Reframing," *Journal of Institutional Economics* 16 (2020): 643–664, doi.org/10.1017/S1744137420000181.

12 Wikipedia, "Limited Liability Act 1855," last modified Jan. 11, 2024, 20:43 (UTC), https://en.wikipedia.org/wiki/Limited_Liability_Act_1855.

THE GOVERNMENT INVESTOR

The next big evolutionary step in the economy occurred after World War II. Presidents Roosevelt and Truman both supported a new vision for funding rapid technological advancement. They wanted to see the United States extend its technological lead, driven by a massive network of research universities and laboratories funded by the federal government. This was an entirely new way of thinking about the interactions among science, technology, and economics. The federal government began setting the conditions to research and develop new technologies. These advances were then turned into revolutionary products and services to fuel the postwar consumer economy—and to maintain our edge over the Soviet Union.

Most people have never heard of General Georges Doriot or his groundbreaking investment firm, American Research and Development Corporation (ARDC). He raised money from a bunch of organizations, pooling it in one fund to make investments in technology companies. It was Doriot, a professor who served as a brigadier general during World War II, who first proved the model could work. It still needed tweaking—the legal structure wasn't quite right—but American investors were figuring out the tools required to finance an entirely new way of building businesses.

For several decades, starting in 1946 with the founding of ARDC, a relatively stable arrangement allowed the technology sector to grow rapidly. The government was the biggest funder of basic research, applied research, and development. It was also the biggest customer for most of the resulting products. A small set of investors, all with deep government experience and connections, supported these new technology companies that made microwaves, radars, semiconductors, and other key products, mostly for national security purposes.

You might know some of the most famous companies that emerged during this period. Digital Equipment Corporation and Hewlett-Packard were two of the first to get started. Fairchild Semiconductor brought together some of the greatest engineers in the country, and many of its early employees went on to found massively successful tech companies and venture capital firms. Intel, America's largest semiconductor company at the time of this book's writing, is an example of a startup founded by the "Fairchildren."

MODERN VENTURE CAPITAL

This government-led era was not sustainable, though. The success of the government's strategy attracted investors who saw the potential for all the game-changing technology. Investors from New York took trips to San Francisco and eventually decided to stay. Their early successes inspired others to set up offices around the area that became known as Silicon Valley. These were small teams experimenting to find out what worked, and eventually they did. This was the era when investors finally cracked the code of the legal structure that's still used today: the limited partnership. Unlike Doriot's ARDC, which was a publicly traded company, investors who organized in private limited partnerships were able to operate quickly and flexibly.

Unleashed with this new legal structure, investors poured money into innovative technologies, driving progress to make things smaller, cheaper, faster, and more reliable—sometimes all at the same time! The massive mainframe computers of the late forties shrank down to minicomputers that a single person could buy, build, and use in the early eighties. This was the time when companies such as Apple, Atari, and Microsoft grew rapidly. They were followed by generations of personal computing and software companies: Sun Microsystems, Dell, Compaq, Symantec, America Online, and many more.

Private investors began writing bigger checks for a wider range of companies. They tended to pick up where the government left off, providing substantial funding for a proven technology that needed to scale up rapidly to avoid missing out on a huge opportunity. This was the modern version of the high-risk, high-reward expedition.

New markets also started opening up. It became clear to investors and entrepreneurs that large businesses could sell products to customers beyond the government, whether to other businesses or directly to consumers. Americans had lots of disposable income and wanted the latest gadgets. Entrepreneurs were building businesses to make it easier and easier for people to get online and find things to do and, especially, stuff to buy.

Startups were not just technology companies either. Venture capitalists began branching out almost immediately. They wanted to know what kinds of businesses and markets would fit into their high-risk, high-reward model. Genentech, a massive biotech company acquired by Roche for $46.8 billion, was venture-backed. So are Starbucks and FedEx. If an opportunity looked like it could grow rapidly, investors wanted to see how they could accelerate that growth and enjoy some of the resulting profits.

Patterns emerged as more and more founders were attracted to the promise of funding for their revolutionary ideas. Silicon Valley investors were able to pick and choose from a large population of creative and talented people. Investors saw thousands of deals and investigated hundreds, only to select the very few they thought could become massive successes. This was a form of "adventure" capital, which was later shortened to "venture capital" and then finally to VC. These VCs got really good at sniffing out winners and letting those runaway successes cover losses from the losers. They also got good at executing strategies to rapidly grow businesses through the maze of risks and obstacles that can easily destroy a promising opportunity.

The most successful entrepreneurs sold their businesses or cashed out after they went public. The newly rich tech founders didn't retire on a beach somewhere, though. They turned around and started more companies or invested in other entrepreneurs. This pattern built on itself, and private investment rapidly overtook public investment in research and development in the nineties. This was the Silicon Valley that generated Netscape, Yahoo, Google, PayPal, Amazon, and eBay.

THE LEAN STARTUP

A set of best practices emerged around startups. These were unwritten rules that people learned to follow through years of experience at high-risk, high-reward businesses. Entrepreneurs and investors knew what worked in those situations; they just didn't have the right language to describe it.

This changed with *The Four Steps to the Epiphany* by Steve Blank (who we will meet in the next chapter). He consolidated all the proven knowledge into one reference guide. It is still the most comprehensive book when it comes to building a startup. Next up came the bestseller *The Lean Startup* by Eric Ries, which popularized the same ideas for the broader business community. Now people all over the world have the words, tools, and techniques—including minimum viable product, Business Model Canvas, and customer discovery—that have been optimized for the early stages of new high-growth businesses.

This series of evolutions is how we arrived at the modern startup economy that you're considering as a community to join. It's a flywheel of people, money, ideas, and hard-earned knowledge, and it is the most interesting and challenging environment you can imagine. Everything about startups is intimidating when you first begin, but you shouldn't worry. The words are different, yes, but a lot of the underlying mindsets and behaviors are the same. The rest of this chapter will give you a quick explainer about the startup world. Everything is oversimplified, but you will get a sense of how the pieces fit together.

There are three basic categories of startups: B2C, B2B, and B2G. B2C stands for "business to consumer," which means the startup sells directly to individual people. Think Uber, Nike, and Apple. B2B stands for "business to business," which means the startup sells to other businesses. Think GE, IBM, and Salesforce. B2G stands for "business to government," which means the startup sells to government agencies. Think Lockheed Martin, Raytheon, and Palantir.

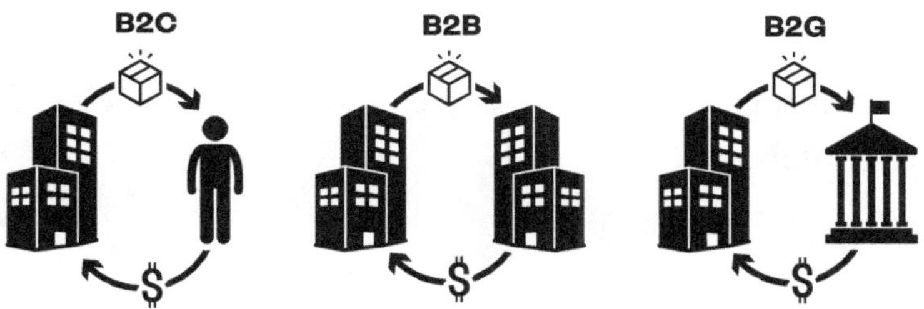

There are three basic categories of jobs inside startups: product, sales, and operations. Product builds the thing. Sales sells the thing. Operations runs the business. Each one grows and becomes more complex as the company grows. Engineering ends up with functions like product management, data science, user experience design, and software development. Sales grows to include marketing and customer success. Operations expands to finance, legal, human resources, and so on.

Product

NCOs are always coming up with creative ways to solve problems, from welding armor to the side of a vehicle to protect passengers from IEDs to figuring out how to reduce the time it takes to complete some pointless task. Companies sell something that solves a problem for their customers. We have to deeply understand the problem to know what to build. Is it a business process that slows down operations or a decision that can't be made because there isn't enough data? People who work in the product arena start by deeply understanding the problem and then using powerful techniques and tools to build effective solutions.

Sales

NCOs are always negotiating. We look for win-win outcomes all the time. This could be haggling to get some additional boots or winter gear from the supply officer, getting your rifle accepted as clean by the armorer, or explaining why the platoon should get out early on weekend liberty. This is based on your ability to get other people to the right answer (which is usually "yes"). That requires listening carefully, getting people to like you, figuring out what they want, and then convincing them that you can help them get it. That is also a great summary of what it's like to do sales at a startup.

Operations

NCOs are constantly finding new ways to drive teams toward specific outcomes, ranging from advising an officer about personnel issues to leading a squad through an exercise. At a startup, operations might involve managing a budget, executing a contract, building a training plan, onboarding new people, or handling administrative issues. When we talk about making sure that shit gets done, we are talking about operations.

THE DIFFERENT STAGES OF A STARTUP

There are three basic stages of maturity for a startup: early, growth, and late. Each phase changes the nature of the work dramatically. In the earlier stages of a startup, the conditions favor Startup NCOs. Life is more chaotic, the rules are less clear, and there are many opportunities to step up as a leader. As the company matures, bureaucracy takes over.

	Early-Stage Startup	Growth-Stage Startup	Late-Stage Startup
Level of Chaos	High	Medium	Low
Size of Company	Small	Medium	Large
Reward for Initiative	High	Medium	Low
Source of Employees	Referrals	Referrals/ Human Resources	Corporate HR

Early Stage

Early-stage startups consist of small teams working eighty-hour weeks trying to figure out what their product is, who their customers are, how the product will actually reach their customers, what to charge for it, and so on. The early stage is chaos. The team is trying to learn as fast as possible because the startup is running out of money. There are thousands of early-stage companies, and most of them do not survive to move on to the growth stage. On the flip side, they do not have mature systems or tools to recruit new employees. Personal referrals carry a lot of weight and will often lead to a job offer if you play your cards right.

Growth Stage

Growth-stage startups are in the process of shifting from a focus on learning to a focus on execution. The fog is beginning to lift. They have found a fit between the product they're building and the market of people willing to pay a lot for it. There are always multiple growth-stage startups competing for the same customers, in addition to the larger businesses that have been around for a while.

The growth stage is a different kind of chaos, because the team is growing rapidly while trying to figure out how to keep ahead of everyone else. There are a few hundred growth-stage startups at any given time. They are hiring rapidly to fill new positions opening up, which is good, but they are also starting to use traditional hiring practices and technologies that make it harder for people like us to be considered for a job. People at this company are likely working sixty-hour weeks.

Late Stage

Late-stage startups are maturing into full-blown corporations. They are usually trying to "go public," which means turning themselves into grown-up companies, allowing investors to recoup all their money plus profit. To go public, a company has to start caring about financial reporting, legal risks, and compliance. Like a snake, late-stage startups are shedding their old skin. Thousands of people with very specific job titles work at these companies, and there is not as much opportunity for career growth. There are a few dozen of these types of startups at any given time.

IT'S NOT THAT COMPLICATED

The world of startups seems confusing, but it's actually very simple. It's all about high risk, high reward, and (hopefully) high growth. There are a few people, called entrepreneurs, who start companies. There are a few people, called investors, who fund them. The entrepreneurs start out with almost no idea of what the company will become; they just have a vision for a better world. They use that vision to persuade intense, hardworking people to join their team. Then everyone works together to figure out how to turn the vision into reality. Some of these early-stage companies make it work. They hire more people to keep up with customer demand, create departments, and adopt policies, and life settles down. Where they once had a startup, they now have a mature business.

You have a chance to grow along with one of these amazing businesses. All you have to do is latch on to one of these high-growth companies while it still has an urgent need for a Startup NCO like you who knows how to get shit done.

THREE

OVERLAPS

Deion Sanders and Bo Jackson are the two best-known athletes in the world who played more than one sport at the professional level. This is a rare accomplishment, even though there have been many Americans with incredible athletic potential. Hundreds of thousands of people are blessed with natural ability and develop it with an intense work ethic. Even so, there are only a few—such as NFL football player and Olympic track-and-field gold medalist Bob Hayes and PGA golf champion and Olympic track-and-field gold medalist

Babe Zaharias—who have been able to dominate in more than one sport.

Michael Jordan and Tim Tebow both tried switching to baseball, but the results were not encouraging. They never approached their original sport's performance level in the new sport. Some people point to Jordan's steady improvement as evidence that he would have been a dominant player eventually, but that is wishful thinking. The world is a better place because he went back to the Bulls to finish out a legendary basketball career.

Athletic ability by itself is not enough to guarantee elite performance. Work ethic by itself is not enough either. Athletes must hone sport-specific skills for years to become good enough to play at a professional level. Here's the catch: all the time spent on one sport comes at the expense of every other sport. Athletes cannot easily repurpose themselves to excel at different movements with different teammates to win a different game.

Fortunately, working at a startup is not like being a standout athlete on a professional sports team. It's more like joining a great team in an intramural league. Teams at this level are looking for solid all-around players. They want foundational qualities: dedication, athleticism, the willingness to work hard, the ability to thrive in chaos, and experience helping a team improve performance. If you have experience playing on a winning team in basketball, a bunch of soccer players will probably give you a chance to prove yourself. They know you are in shape and can play well with others.

The approach is similar for startups: it's okay if your experience isn't well-rounded or you didn't go to all the right schools. You don't need to have the same level of polish and credentialing as you would at a larger business. You don't need to check all the boxes on your résumé. These businesses are desperately searching for high-quality people who can thrive in a specific kind of high-growth environment. They want someone like you—they just don't know you yet. Your job is to help them realize that they've been looking for you all along.

Think back to the last chapter. A startup pursues massive rewards in a high-risk environment. The founders are just getting started. There are a million problems to solve all at once, but there is little process or structure to fall back on. The culture, if you can call it that, is whatever people do when they're running around trying to get work done. The business faces intense pressure to show impressive results on a tight timeline. New people join without much direction. The whole situation is stressful, messy, and overwhelming.

What does this situation sound like? A lot like life as an NCO in the military, especially as part of a small unit. And that is an incredibly important point to internalize.

The entire book flows from this one key insight: A startup and a small unit operate in similar ways. If you can thrive in one, then you can thrive in the other. You can actually be a multisport athlete, because the two sports turn out to be very similar. The military got you in shape for a different sport without your realizing it!

STARTUP-MILITARY PARALLELS

This is a big idea, and it is worth unpacking more fully. Right now, you may think startup life and NCO life appear to be very different from one another. There are many cosmetic differences between them, obviously. The organization is different, the mission is different, the work is different, and the cultures will be different. If you

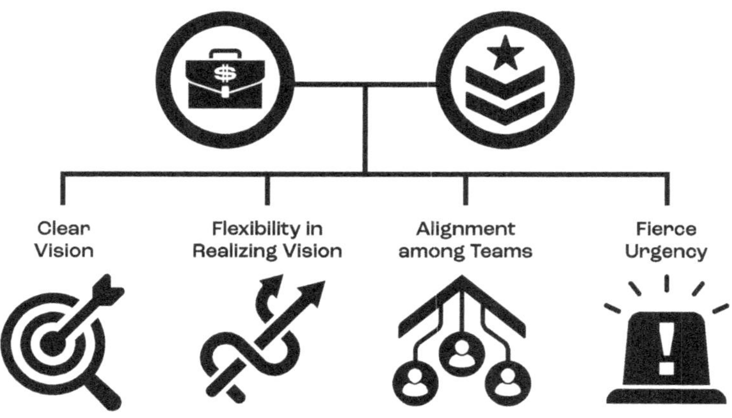

strip away the surface level and get down to the fundamentals, though, you will find out that they are very similar. They are high-performing teams trying to win, one at war and the other in business.

Consider the leader. For a military unit, that is the commanding officer. For a startup, that is the chief executive officer. CO versus CEO. In both cases, the leader has a compelling vision for what they want the team to do. This mission is the "why" of the team, its beating heart. The leader spends a lot of time talking to the rest of the team about that vision and why it matters.

On its own, however, a vision is just...well, a vision. It is not a reality. A ton of hard work lies between the vision as it exists in the head of the leader and the team's ability to operationalize that vision in the real world. Without NCOs, a CO would quickly discover that the unit degrades very quickly. Startup CEOs are desperately searching for people who serve that same function: taking the vision and figuring out how to make it a reality. The typical response at startups is to push this responsibility onto anyone and everyone. Why? Because the CEO doesn't have an NCO.

At startups burning through their cash, CEOs are on the clock. They need to discover—and quickly build—a real business that fits within the startup's vision. There are specific business targets that CEOs need to hit. If they miss enough of those targets and can't get back on track, then a replacement will come in or the startup will shut down. The company will either grow quickly or cease to exist. There is no middle ground, and staying on that rocket-launch trajectory requires a monstrous amount of work by the team.

Planning and working insane hours are both key parts of life in uniform. Leaders in the military love to plan. They have all kinds of fancy processes and checklists and acronyms to describe the approved ways to plan out every conceivable activity. And yes, planning is an indispensable part of any operation, even if the team doesn't follow the plan at all. NCOs know the pressure on small-unit leadership, especially at the tactical level. We are often forced to make decisions that change "the plan" in important ways. That is the origin of the granddaddy of acronyms, SNAFU: situation normal, all fucked up.

THE PIVOT

Like commanders in the military, startup CEOs also have plans. They will likely be based on extensive experience in the industry, possibly drawing on academic research and hundreds of conversations with experts and mentors. These plans will probably sound pretty good when you hear about them. They will also look good on PowerPoint slides. Some of your coworkers will even get suckered into thinking that the plans will somehow work. An NCO knows better: these plans will not survive first contact with reality.

Silicon Valley is full of legends about companies that changed direction completely, in what are commonly known as *pivots*. They are very common in the early days of a company. Twitter started out as a podcasting company. Instagram was a check-in app. YouTube was a video dating site. Slack was a gaming company.

When they began, none of these CEOs knew that the initial plan was wrong. They had to start building and then make tough decisions based on real-world feedback. The burden then fell to the employees to build something that was actually valuable to customers. That is the "finding the business within the vision" phase that is so exciting and challenging at the same time, and it's key to the mission of early-stage startups. The essential skills required in that phase are learning quickly, committing to decisions, and maintaining high performance despite a grueling workload.

STARTING AT A STARTUP

I was introduced to the insane demands of startup work early on. I joined a startup after graduating from college, and my onboarding coincided with that of the first senior technical hire. He reviewed all the code in his first few weeks and discovered that the software developers had made some pretty serious mistakes. He had to rebuild the back end of our product using a different programming language. This meant that we had to move a whole bunch of unstructured data from the old database to a new one. There were fancy ways to do this, but they were expensive. A few nontechnical people like me were hanging around, though, so the CEO told us that we had to do it manually. At the rate we were growing our users, the problem was getting worse every day, so we had to hustle to get it finished.

I did exactly what any NCO would do. I asked a few questions of the CEO, COO, and CTO to make sure I understood why we were doing the work. Then I went into the bathroom, calmly shut the door, and cussed loudly for a few minutes. I came out and—again very calmly and as if nothing had happened—started planning out the work with the rest of the team. We drew a diagram of the workflow on a whiteboard and assigned a person to each section. We did some basic calculations to figure out how much work we had to do each day in order to knock out the whole thing in two weeks. We canceled our meetings for that time period, rolled up our sleeves, and got going. I gave the three leaders updates every day and made sure to check in with each team member whenever it looked like they were hitting a wall.

By the end of this project, I was worried I was going a little crazy. I stared at spreadsheets so much that I saw columns and rows when I closed my eyes to go to bed at night. And sometimes I didn't even get to go to bed! We had to pull multiple all-nighters to get everything done. To meet the deadline, the CEO even paused most of *his* meetings, except for the ones with investors. (*Side note*: You do not ever cancel meetings with investors.) We barely made it, finishing on the last day with no time to spare.

This is a good example of an early-stage startup problem. One day we discovered that something was seriously wrong that threatened the business. We then had two weeks to do what was realistically two months' worth of work. If we didn't get it done, we would have to shut down, plain and simple. Ironically, we ended up shutting down four months later, but for an entirely different reason. Such is life at an early-stage startup.

FEAR OF FALSE FAILURE

After reading this far, you may be wondering if this path is worth it. After all, what's the point of putting in all this effort if the first company you join shuts its doors four months later?

Look, many people are afraid of startups shutting down. They fear losing their job. This is understandable—but also completely misguided. Startup NCOs will always have job opportunities when they want them. What appears to be a bug is actually a core feature.

Startups are created and shut down all the time. This dynamic leads to a unique feature of the community: people are always changing jobs. Someone working at Google might leave to build a startup. If it fails six months later, that person will find a job at Salesforce or Amazon in a few days, or they will decide to try another startup if they have the right cofounders, an awesome idea, and enough cash in the bank. This kind of rapid rotation through a series of jobs is normal. In fact, it is encouraged so that people can find the opportunities that are right for them.

Reputation is everything in the startup world. People who get shit done will always have jobs waiting for them, sourced from a wide network of former colleagues who ended up at other companies. Do not worry about whether your first, second, or even third job is the one that works out. Focus on creating value and making friends (which we will discuss in more detail later), and you will survive just fine. In fact, you will thrive.

Remember, the startup game is all about survival of the fittest. CEOs don't have the luxury of focusing on all the "nice to have" traits that more traditional employees can bring to the table. If a CEO has to choose between people who know how to grind it out as a team and individual achievers who fall apart when things get rough, then guess what? A good CEO is going to choose the grinders. That's where you come in. You have to convince them that you know how to lead a team through the grind to turn their vision into reality. You have to convince them that you are a Startup NCO.

FOUR

STARTUP NCOs

Steve Blank is the archetypal Startup NCO. In case you haven't heard of him, here's a short summary. He grew up in New York and then enlisted in the Air Force during the Vietnam War. He went overseas to work on electronic equipment, learning about some of the most impressive technologies of the time. After leaving the service, he moved to Silicon Valley in 1978 and began working in startups. He discovered a talent for marketing and rapidly moved into key leadership positions. By the time he left his final startup, E.piphany, it was ready to go public on the Nasdaq stock exchange. He was already rich, but that company made him a multimillionaire.[13]

Steve "retired" and became a teacher, author, investor, and advisor. He's written multiple best-selling books, developed the world's leading curriculum for entrepreneurs, and created the American government's premier training program for researchers and scientists. Despite being a college dropout, he has taught at some of the world's best universities and given several university commencement speeches.

Steve showed up in Silicon Valley with no credentials, no network, and no degree. He had something better than all the cosmetic stuff that fills up most people's résumés, though. He was that special kind of person that high-growth companies desperately need: a Startup NCO.

I think Steve is a particularly strong model of a Startup NCO because he began with no connections or formal business training but went on to successfully draw on his experiences and strengths in a series of demanding startup jobs. Starting out, he was curious about the technical systems used in signals intelligence. Fortunately, he had the opportunity to read through manuals, play with the gear, and talk to experts. When he got out of the Air

13 Wikipedia, "Steve Blank," last modified Sept. 15, 2024, 00:13 (UTC), https://en.wikipedia.org/wiki/Steve_Blank.

Force and came to Silicon Valley, he immediately set himself apart by putting these same behaviors into practice. He has shared how much his enlisted service helped him. He learned creative problem-solving, great responsibility at an early age, and a broader perspective on life.[14] He also set a high bar for how much work he was capable of doing. All this, applied over decades, turned him into an indispensable business leader.

Steve was the first Startup NCO to break into the world of high-growth technology companies just as the industry was taking off in the late 1970s. He was promoted rapidly, taking on ever-increasing responsibilities as he went from one early-stage startup to the next. Despite these changes, he kept practicing the same fundamental behaviors: reading, listening, thinking, learning, and leading. His life is a testament to what the rest of us coming out of the military can achieve. We can become Startup NCOs just like him.

The process is not easy, but no process of transformation is easy. It takes years of training and experience to develop someone from a junior enlisted service member to an NCO. It will also take years to adapt your military

14 Bill Murphy Jr., "5 'Military Secrets' behind Steve Blank's Entrepreneurial Success," *Inc.*, June 6, 2013, https://www.inc.com/bill-murphy-jr/5-military-secrets-behind-steve-blanks-entrepreneurial-success.html.

mindset and skills to the startup world, but don't be intimidated by the length of the journey. The best time to start this transformation is today. There will be many important milestones along the way, including the incredible feeling of getting that first job at an awesome startup.

Before beginning that process, however, we should walk through the principles that you must understand if you want to become a Startup NCO. These are not specific techniques or tactics that will guarantee success. They cannot be implemented one at a time either. These are ways of thinking and behaving that build on your military experience. You are not starting over but rather adapting what you already know to a new environment with new challenges and opportunities.

COMPLEMENT THE CEO, ENABLE THE TEAM

As NCOs in the military, we adapt to the CO of the unit. That CO's goals, quirks, and whims all factor into how we operate. NCOs are the crucial link between the leader and the bulk of the team. The same is true in a startup. A Startup NCO understands that a business is still a group of people. Strategies, processes, and tools all have their place, but they only create value when people adopt and implement them. Thus, an important aspect of being a Startup NCO is paying close attention to the way the CEO leads the team. Knowing the CEO, we can adapt our approach to that person to ensure the rest of the organization operates at a high level. In practice, this means emphasizing the CEO's strengths and minimizing their weaknesses.

ADVISE, THEN EXECUTE

A key job of NCOs is advising the CO. We are the only part of the team that bears the responsibility of speaking truth to power. That means calling out the leader when they are about to do something stupid (even though it sounds smart when it's explained on a PowerPoint slide). This is an important role that evolved over time as the nature of our missions and structure of our organizations changed. Regardless of the details, good officers know they need sanity checks. They need to hear when what they are proposing simply won't work.

The flip side of this advisory role is to then become the champion of the plan once the CO says it's time to execute. The same is true of working with CEOs. There is a time to debate any difficult decision in private. Then the team commits to a decision and moves out. Startup NCOs know that other people watch them carefully to gauge enthusiasm and commitment. We are, therefore, the most enthusiastic and committed out of everyone in the company. The CEO deserves nothing less.

CUSTOMERS FIRST, PEOPLE ALWAYS

The saying in the military is "Mission first, people always," and any unit with a healthy culture puts that principle into practice. Startup NCOs operate in a business context. Unlike the military, there is an overwhelming need to make enough money to pay the bills. How do businesses ensure they make money? By taking care of customers!

Sometimes this gets out of hand, however, leading to the neglect of employees. The private sector doesn't have the same expectations of leaders to take care of their people as we do in the military. This oversight is tragic and must be avoided at all costs. Startup NCOs know the importance of making coworkers feel like they matter. Employees are not expendable assets or cells on a spreadsheet; they are people. They deserve to be treated with respect, especially when the business has to make tough decisions about who to keep and who to let go.

LEAD BY EXAMPLE, ENFORCE STANDARDS

In the military, we are judged by our actions rather than our words. It's one of the few sectors of society where concepts like honor don't feel old-timey or out of place. You and I are incredibly fortunate to have been molded in that environment. We know what it takes to instill and maintain these values in our teams.

Startup NCOs need to bring that same level of diligence to the standards that are critical to the company's growth. People need to care—and to know that others around them care too. Whatever the company's values, they need to be translated into a set of behaviors. This helps employees understand "how we do things around here" without being given a set of rigid rules to follow. Startup NCOs find effective ways to bring values to life. There cannot be any gaps between the way the company talks about itself and the behaviors we tolerate on a daily basis.

WATCH ONE, DO ONE, TEACH ONE

NCOs are masters of creating learning organizations. This is a fancy way of describing a culture that rewards teaching, coaching, and mentoring at all levels. You were probably taught something along the lines of "you should know how to do every job below you, as well as the job of the person above you." NCOs have to rapidly onboard new people by getting their hands on the systems and tools they need to get the work done. There is no excuse for laziness when it comes to developing people.

Startup NCOs have to achieve the same thing, with the added complexities that accompany working at a dynamic, fast-growing business. Startups are not usually ready to standardize processes until they've been around for a while, so Startup NCOs carefully judge what's worth embedding into the culture. These are relatively small decisions, made almost daily, that quickly add up to important aspects of how work gets done.

SIMPLIFY

My favorite (unofficial) Marine Corps slogan is "Officers: making simple stuff complicated since 1775." NCOs are the layer between officers and junior enlisted members and, as such, quickly learn that "perfect" is the enemy of good. A lot of plans make sense when a group of officers stand around drinking coffee. But what happens when you take one of those same plans and try to explain it to a group of twenty-year-old Marines? The whole thing falls apart.

This gap only gets worse in the private sector. Smart people come up with plans that have no connection to reality, an especially big problem at startups. Many CEOs spend a lot of time selling an ambitious idea to investors and customers, forgetting that someone actually has to make it a reality! Startup NCOs are incredibly good at taking a complicated plan, distilling it to the essential points, and then helping a team understand how everything fits together and, most importantly, which specific things to do right now.

LEARN HISTORY, THEN TEACH IT

The military has a difficult job. It must take hundreds of thousands of Americans each year and turn them into functional service members. These recruits must understand how they fit into a specific branch of service. Each branch has a rich history, full of people doing impressive things. Learning these stories helps people see themselves as a part of something greater, which is critically important to achieving the mission. Storytelling lays a foundation that helps teams surpass limitations and achieve amazing things together.

Just like the history of the Army or Navy, there are thousands of inspiring stories of startup founders who built something out of nothing by taking massive risks that paid off—everyone from Christopher Columbus to Steve Jobs, and you will hear many more in the years to come. These stories form important connections to our past, and Startup NCOs know how to weave them into conversations with their teammates. This history connects onboarding employees to a community that can inspire, challenge, and sustain them in both good times and bad.

YOUR JOURNEY BEGINS HERE

Throughout Part I of this book, I have tried to make the case for good NCOs to join startups:

- The introduction provided a mix of personal examples and historical evidence.
- Chapter 1 traced our shared lineage as NCOs, including our unique role in the professional, all-volunteer American military.
- Chapter 2 described the evolution of high-risk, high-reward ventures over hundreds of years. This model funded naval expeditions to discover new land and went on to industrialize nations. Now, it funds promising new businesses.
- Chapter 3 highlighted many of the overlaps between the pressures faced by an NCO and those faced by startup founders. Regardless of the vision or the mission, a team needs to confront reality to figure out how to carve out a successful niche for itself.
- Chapter 4 clarified some of the high-level Startup NCO principles that will be especially helpful when

working with a CEO to deliver on that compelling startup vision.

Part II assumes that the arguments presented in these chapters have persuaded you to dive deeper. You are interested enough in the potential for a fulfilling, lucrative startup career to learn more about the journey. I will start by sharing some more insights from my own journey and then discuss three key areas for personal and professional growth.

PART II

BECOMING A STARTUP NCO

I had no idea what I was doing when I joined my first startup, which was a standard-issue growth-stage startup. It had been around for a few years, and there were already several hundred employees. Another Marine who already worked there met me for a coffee, and we spent about forty-five minutes together as he walked me through the business. We identified a place where I might be useful in supporting one of his employees who was working on a federal government project. They were part of a small team that fit into the sales function of the company.

ROUND 1

My boss, the Marine's subordinate, was a civilian with an MBA who had worked in corporations all his life. This was also his first startup. I was ahead of him, though, because I knew that I was starting from scratch. He acted like he knew exactly what to do, approaching our project as if we worked for a regular company, not a chaotic startup. He put his head down and started grinding away on the initial version of a project plan.

The startup was in San Francisco, so I had to take the train up from Stanford, where I was finishing my degree. I was so ignorant that I assumed it was okay to ask a lot of questions. I'd show up at the office, and my boss and I would resume the conversation that had never really ended the day prior. At first, my questions related to the project. Eventually, they started to probe the limits of his knowledge about the startup. What could I do to be successful there? How would we know if the project was successful? How did our work roll up into the larger business? How did the company make money?

These questions were obviously very annoying to him, especially after a few weeks working together. However, he needed someone to do a bunch of his BS work, and I was the only person who had been hired onto his team. So, he humored me. The guy didn't really have a choice. Who else was going to spend hours tweaking his slide decks and cleaning up his spreadsheets?

Most of the people I met there were MBA types who reminded me of the officers I knew in the Marines. They were confident, well-spoken, and mild-mannered. They checked all the boxes you would want in a generally competent person to join the team. They could research and write a great white paper or build a solid presentation in a slide deck. They were high-performing general-purpose knowledge workers (we will come back to that point later on), but they did not have practical knowledge regarding how this high-risk, high-reward business could become a great success.

My questions got more in-depth as I picked up the basics of the business. Pretty soon, my boss didn't know the answers to most of what I asked, so he started repeating my questions to his boss. That worked for another few weeks, but eventually we reached the point where my boss's boss did not know the answers to my questions either. This was a fascinating lesson for me. Their knowledge was shallow but dressed up in fancy language that hid their underlying ignorance about the fundamentals of the business.

The work didn't last long. I was a contractor, not an employee, so I was easy to fire. The company decided to scrap my position after one of my projects wrapped up and there was nothing immediately available to justify my monthly retainer. I was not crushed, though. I now had six months of experience under my belt and was getting a feel for startup life.

Fortunately, another Marine—hopefully you're seeing the pattern—reached out around that time to see if I would be interested in joining a different company. Every time a door closed, several more opened up. The fear of "losing my job" was intense when I first finished college, but I never actually experienced a problem. My reputation in the startup community ensured that someone was always interested in bringing me on to a great team pursuing a worthwhile mission.

ROUND 2

My next startup was just getting back off the ground after a first failed product launch. This was a stereotypical early-stage startup, with a team of fewer than a dozen people, a working product that had a lot of bugs, and almost no revenue. My buddy was the most recent hire, brought in to be the chief operating officer (COO). He knew I was opinionated, a pretty good researcher and writer, and sharp enough to learn on the job. He also liked that I had a bit of startup experience, which included a referral and recommendation from the Marine who had hired me previously. After a few interviews, including one with the lead investor, I was hired.

All of a sudden, I was a real employee at a real startup, even flying out to the headquarters in Scottsdale, Arizona, and staying in a corporate apartment. The team worked hard—pulling all-nighters more than a few times—and played hard, unleashing a small amount of hell at the apartment where we were packed in like sardines. It was a roller coaster from week to week, with high highs and low lows as we tried to hit targets for growing users and revenue.

Again, the work didn't last long. Seven months later, waiting on two weeks of back pay, I was told that the company was shutting down. The multimillionaire lead investor who had helped recruit me decided that the business wasn't going to work. When the CEO asked for another check to keep the company going, the investor said no. This meant that the junior employees like me would not be paid for the work we had done that month. I was now out of work and stuck with a lease on a Palo Alto apartment that I could not afford.

The startup dream had crumbled again. But, as before, something else came along. And this time, I was starting to see that I would always have some kind of opportunity available through my network. My growing reputation as a reliable teammate for early-stage startups offered me a steady flow of interesting jobs. I was also fortunate enough to find a way out of my apartment lease and somehow tricked my brother into letting me stay in the extra room at his place in San Francisco. My expenses returned to a manageable level, and I could focus on finding the next opportunity.

ROUND 3

It's hard to describe the cofounder of my next company. Joe Felter is a Special Forces guy who saw action as early as Panama in 1989. He also earned a PhD, and we met when he was guest lecturing in a class at Stanford. Joe is a cross between GI Joe and the Absent-Minded Professor, and he is one of my favorite people. He reached out to me because an Army colonel named Pete Newell was visiting campus. Pete wanted to see a dozen groundbreaking startups that could help him solve some problems in Afghanistan. This was 2012, and the mission-critical challenge was improvised explosive devices (IEDs). Our job was to find startups with technologies that could be repurposed to solve the IED problem.

Joe and I were already friends, and Pete fit right into the group. We trusted each other from the first time we met. It quickly became clear that the three of us shared a mindset, set of values, and burning desire to get after military problems with commercial technologies. That was all it took. BMNT, my third startup, was born, even though it took over a year before Pete retired from the Army and we could legally found the company.

Joe, Pete, and I were joined by Brendan (another Startup NCO who eventually left to build a company on the East Coast). We decided that the business we wanted to build would not focus on growth at all costs, so we never raised any outside money. In Silicon Valley, this approach is known as bootstrapping—pulling yourself up by your own bootstraps, as the expression goes. The upside is owning the company and not having to listen to what anyone else says. The downside is only having the money paid by customers. There is no upfront investment of cash to infuse resources when a bootstrapped company first starts out.

Despite the lack of venture capital, we've grown into a business of almost a hundred people, with offices on both coasts and partnered companies operating in the UK and Australia. During this same period, I tried other businesses as well. Two weren't successful, but one was profitable enough that it provided my wife and me with the down payment for our first house. You have to keep your startup skills sharp even as you adapt to the needs of a growing company.

This all probably sounds pretty chaotic—or, at least, I hope it does. These are real-life examples of what it feels like to be a Startup NCO! I spent over two years trying to break into this world. I met hundreds of people, interviewed with dozens of companies, and ultimately cycled through three different startups. This level of turnover is not normal for everyone, but it happens. No one promises a smooth ride in the military. No one promises a smooth ride at a startup either.

Even though my main job is running current operations for BMNT as its COO, I still test out new ideas on a regular basis. I want to find the next big thing that will quickly grow into our biggest source of revenue. The urge to build never left me. It probably won't leave you either.

RIGHT WHERE I WANT TO BE

The early stages of company-building are precisely where I want to be. Those phases are where I find the greatest satisfaction. The environment consists of only the founder's vision, a few people, and a complete lack of structure or process. It's controlled chaos with a high-trust team that is struggling to stay alive. Those conditions are where we as NCOs are best suited to thrive, where it is okay to be rough around the edges or even (gasp!) not a graduate of an Ivy League school. The ability to do hard work in tough situations gets us in the door, and our relentless drive to learn and adapt allows us to thrive over the long term.

None of these achievements comes easily. Remember boot camp? It took weeks of training before we could even march in step with other recruits. After that, it took months of training before we became proficient in our Military Occupational Specialty (MOS). Even after being trained, it took years of experience to become truly good at the work, no matter the MOS. Transitioning to startup life is similar, so expect a similar timeline.

Anything worth doing—and anyone worth becoming—takes years. We are changing our identities, remolding our sense of self. We are slowly learning to see ourselves as people who can stand toe to toe with peers in the high-stakes startup world. We are developing a strong sense of belonging in a new community full of entrepreneurs and early-stage startup workers.

The next three chapters will walk you through the three elements of this transition: first, what to keep from your time in the military; second, what to forget from your time in the military; and third, what to learn during your early-startup career.

FIVE

KEEPING

Bill Perry enlisted in the Army during World War II but didn't deploy until the end of major combat operations. He was stationed in Japan, where he saw some of the devastation from the nuclear bombs that the United States had dropped to end the war in the Pacific. He personally met survivors of Hiroshima and Nagasaki when taking part in stabilization operations.

When Bill was discharged, he went back to college on the GI Bill, earning two degrees from Stanford and eventually a PhD from Penn State. He started a few companies during the fifties and sixties before returning to government in the seventies. After holding several senior positions, he headed back to Silicon Valley to run an investment firm. He was eventually called back to service by President Clinton to become the secretary of Defense. As of this writing, Bill is still deeply involved at Stanford regarding issues of technology and national security.

When Bill was secretary of Defense, he was one of the most powerful people in the world. He enjoyed an unbroken string of professional successes. Yet somehow, he was also one of the most down-to-earth people you'll ever meet, and a great leader. The distant memories of this former NCO fundamentally shaped how he viewed the world.

Bill showed up in little ways that, if you asked him, he would credit to his Army basic training. For example, Bill was known for taking out his own trash, even as secretary of Defense. Many people thought this was strange. For us NCOs, however, it makes sense. Of course you take out your own trash. Who else would take it out for you? This may not seem like an important point, but it is actually essential. A single behavior, taking out the trash, relates to a core part of Bill Perry's identity and his formative years in the Army. His decades of professional success were not accidental. They resulted from the consistent application of basic principles that he picked up as a young enlisted soldier.

You also have the same principles, values, and behaviors. They were drilled into you the same way they were drilled into Bill. He worked hard to become a Startup NCO. So can you.

As I have already discussed, it's initially hard for many of us to understand our own value to a startup. We're not yet familiar with the intense pressure and chaotic environment. We also struggle to see the limitations of the highly educated, credentialed people around us. It's easy to dwell on our weaknesses instead of proudly highlighting our strengths. This distorted view limits us. We have to see ourselves clearly if we hope to understand the value we bring to the table as NCOs.

The startup environment is full of activity and confusion. There are predictable patterns of behavior that drive performance in the midst of chaos, from servant leadership to effective communication. CEOs are constantly scouting talented folks who can help them build a winning team. They want people with the qualities that have been hammered into us during our time in uniform. These leaders see the immediate positive impact when a Startup NCO comes into the business. Why? Because we develop and implement a series of fancy new strategies to take the business to the next level? Hardly. Startup NCOs create value by applying the same time-honored, battle-tested principles that we used in the military. That is why we begin Part II with an entire chapter focusing on the parts of yourself that must be kept at all costs.

KEEP YOUR HUMILITY

NCOs are not the top of the hierarchy and never will be. We do the thankless work that keeps the military humming along. We do not come up with ideas, but we do ensure they are understood and executed. We do not come up with the values, but we do ensure they are practiced every day by every person under our charge. It takes humility to set boundaries on our role, focusing primarily on the team's execution of the leader's vision.

What personality type do CEOs struggle with? People who want to come up with new ideas instead of executing the existing ideas from the CEO. These Good Idea Fairies sprinkle dust everywhere and clog up the business. Do not be one of them. Know your circle of competence. Stay humble.

KEEP YOUR RESILIENCE

The military forces us to be resilient, to bounce back from adversity. We have fallen many times and have always gotten back up. There is really no other choice. We were told stories about resilience from our first days in the military, such as the citations of Medal of Honor recipients from past wars. We also watched the people around us demonstrate resilience. They accomplished the seemingly impossible without complaint, then turned around the next day and did it again. Finally, we experienced a warrior culture that prizes resilience. We soaked up these influences, learning to think of ourselves as resilient. Over time, we also learned to foster that sense of resilience in others. Many civilians lack any experience like ours. They will quickly learn that you are a different sort of person and start to follow your example.

KEEP ADAPTING

Plans are always changing. The conditions in which we operate are always changing. When things change, NCOs have to figure out what to do at the operational and tactical levels. Adaptation is a constant companion, a continuous feature of our lives. We know other people will screw up in the planning stage, and the consequences will trickle down to our level.

This situation does not happen to the same degree in most other organizations, so adaptability is not developed to the same degree. Many civilians will roll their eyes and complain about changes rather than figuring out what to do and getting after it. We must keep the mindset that anticipates change and prepares our teams for it. Doing so immediately sets us apart from others when the CEO notices who actually gets the job done.

KEEP COMMUNICATING

Former Secretary of Defense and Marine Jim Mattis instructed his leaders to ask three questions of themselves:

1. What do I know?
2. Who needs to know?
3. Have I told them?

This is another way of saying, "Pass the word." Surprisingly, there does not seem to be the equivalent of "passing word" in many companies. There is Slack. There are all-hands meetings. There are email blasts. These techniques and tools all help to generate information, but they do not pass the word. They do not intentionally spread critical information via a trusted network in such a way that people actually understand what is going on and how it affects them.

As NCOs, we learned to ask the Mattis questions. In the dynamic environment of a startup, people prize trusted, timely information. Anyone who provides it will quickly become indispensable.

KEEP SOLVING PROBLEMS

NCOs always have problems to solve. It could be something as simple as an administrative issue or as serious as a medical problem that needs immediate attention. It might be a lance corporal who sold their gas mask for some extra drinking money between paychecks. (I wish I could say that problem has never happened.) We spend years caught in the middle between junior enlisted folks and officers. That is, pardon my French, a shit sandwich that no one wants to eat.

Fortunately, we've gained valuable perspective on how to solve problems in the middle layer of an organization. This turns out to be very useful. In particular, we already have a finely tuned sense of which problems need to be solved through formal channels versus pulling someone quietly to the side to let them know they screwed up. We match the problem to the solution in a way that reinforces the organizational culture. CEOs crave this behavior because it protects against the creeping tendency to slow down great teams with unnecessary process and restrictive policies.

KEEP FOCUSING ON THE TEAM

In the military, everything is about teamwork. We do every activity as a team, from training exercises to filling out paperwork. We drive in convoys. We have battle buddies on liberty. We are always checking rosters. We even do piss tests as a group! These experiences train us to think about the group first rather than ourselves.

Civilians who lack those same experiences immediately appreciate our group orientation. They appreciate the instinct to care for the group no matter what else is going on. Startup NCOs put the team first in such a natural way that we aren't even aware of it. Coworkers respond by mimicking the behavior, which helps the team perform at a higher level and adds another layer of positive culture that helps the whole business perform at a higher level.

PURPOSE-BUILT TO THRIVE IN CHAOS

These Startup NCO qualities and behaviors are incredibly valuable. No other institution in the country develops these traits in people from such a young age. We are purpose-built to thrive in chaos, which is a good functional description of life working at a startup. That does not mean you can just stroll into the office like some kind of demigod, however. Not every behavior from the military is worth keeping. We can appreciate the good while recognizing there is also bad. That is the subject of the next chapter.

SIX

FORGETTING

In the summer of 2007, I was at Camp Pendleton doing predeployment training before my unit headed to Iraq. The command was trying to knock out the entire set of annual training requirements in one day, back to back to back. It was death by slide deck. The audience was all combat arms, which at the time meant it was all men. I sat with a bunch of other Marines waiting for the next brief, which was sure to be as boring as the last one. A professional-looking woman walked up to the podium and dove right into her slides on equal opportunity and sexual harassment.

Within two minutes, sidebar conversations had started. The volume steadily increased, and pretty soon this poor woman had to raise her voice despite speaking directly into a microphone. It was clear that no one cared about what she was saying. The hundreds of Marines filling the auditorium collectively communicated that they did not value this training. Leaders in the back of the room did nothing to intervene.

This was one of thousands of my early experiences as a young infantryman that conveyed a simple message: women aren't equals. Of course, they might have been equals in some abstract political sense—they get to vote, don't they?—but not in any practical way. Disrespecting them didn't seem like a big deal. It was also perfectly fine to gaff off this mandatory training designed to teach young men the basics of professional conduct in a modern work environment.

Let me pause here to say how much I love the United States Marine Corps. I love our ethos. I love our warrior culture. I love our mission. I love Marines. I wouldn't put on any other uniform if you paid me a million dollars. That said, I do not pretend the Marine Corps is perfect. It is a flawed human institution with a checkered past. So are other branches of service.

The military's emphasis on history can cut both ways. History can be—and often is—used as a justification for ignoring important and necessary changes in American society. What one group of people in a previous era considered normal will be viewed differently by people today. A gap emerges if we don't somehow reconcile these beliefs. The wider this gap gets, the easier it is to frame things as "we" and "us" versus "they" and "them." "We" do things a certain way, the right way, while "they" do not. People in the military have to be wary of allowing this divide to lead to sloppy thinking and bad habits that will cause problems when they are trying to get—and keep—a private-sector job.

Sex and gender offer one example of this gap. There has been meaningful progress in providing opportunities for women in the military to join direct-combat organizations so long as they meet the same high standards as their male counterparts. More women are in top leadership roles across the military than ever before. And there is growing recognition that the military can't meet its recruiting and retention goals if it allows outdated cultural practices to continue. Otherwise, some of our best and brightest will leave for the wrong reasons.

But the military is still, well, the military. It's mostly young men. Young men do and say a lot of dumb stuff, especially enlisted folks when they're all packed together for long periods of time without a lot to do. That kind of environment doesn't set you up for success in the outside world. It's easy to adopt certain behaviors, especially ways of speaking, that would shock civilians with no frame of reference for what they were hearing. There are specific examples later in this chapter. For now, keep in mind the basic idea that you will have to let go of some old destructive behaviors to make room for new productive ones.

LET IT GO

Unlearning destructive habits is the first challenge faced by an aspiring Startup NCO. Don't start by trying to add new knowledge or skills, or to build out your network. Start by removing, by forgetting, by peeling away layers of your military experience that are no longer necessary for you to thrive on a new kind of team that you've never experienced before.

Startups share a lot of similarities with military organizations, as we've already discussed. The dynamic missions, the high operational tempo, and the ambiguity will all be familiar. However, there is a fundamental difference in the social environment. It is much more difficult to forge a high-performing team. Why? Because your teammates will not share a common identity, either with each other or with you.

Until you leave the military, you can't understand the full impact that shared identity has made on you and everyone around you. The civilian world is a free-for-all. As civilians, we identify with all kinds of weird things. We may base our identity on the college we went to, our religion, our political party, or our racial group. Others may be really into a favorite sports team, or being a dog owner, or some obscure rock band. How do you get a "MAGA person" to work effectively with a "Dua Lipa person"? This might sound like a silly or trivial challenge, but it is critical to building a world-class team.

Imagine yourself starting a new job. You're staring at your new colleagues, realizing that you have almost nothing in common. You have almost no way to build relationships with them, except if you happen to get along. You're stuck as an "other" if you can't figure out how to meet people where they are, and you may have zero in common to help you do that.

FRIEND REQUESTS

It took me years to figure out how to make friends with civilians. I did okay at community college by joining a student group focused on academic performance. I liked people who came from a bunch of different backgrounds but shared a goal. In a sea of mediocre students, we were all there to learn. We were the hardworking nerds sitting in the front of the classroom. That made it pretty easy to find ways to work together. I did not have friends, but I could find people to join me for a study group.

That all changed when I transferred to Stanford University. I had just returned from a deployment to Iraq six weeks earlier and was ready to get back to school. I had no idea what I was in for. At community college, I could single out hardworking nerds and try to make friends with them. Now, *everyone* was a hardworking nerd, even the football players! The student body was aesthetically diverse, of course, but it was 99 percent kids under the age of twenty-two who had no life experience. As a Marine, I was a freak. As a twenty-six-year-old Marine, I was an old freak.

I tried to make friends a few times but mostly struck out. The gap was just too wide. Almost every conversation petered out after I mentioned my time in the military. I lasted about a year and a half at Stanford before throwing in the towel. I heard about President Obama's surge in Afghanistan, learned about the central role the Marines were playing in Helmand province, and decided I wanted to go back in. I talked to my dad (who could see I was serious about this), notified Stanford that I was dropping out, reenlisted, and jumped on another deployment.

For some reason, it was easier for me to put the uniform back on and go to war than to stroll through a beautiful campus and do a few hours of homework each night. It wasn't the classes; I had enough consistent performance as a college student to know I could hang with my peers.

There were two big issues: I felt like an outcast, and the work seemed pointless. The outcast part is self-explanatory. I wasn't making friends except with the tiny number of enlisted vets on campus and with a few ROTC cadets. I felt no connection to the students around me. I had nothing to say to them, and I imagined the feeling was mutual. Then there was the work. It was hard to stare at a civil-engineering problem set when I knew the Marine Corps was gearing up for a big push into Marjah. So, I had no friends and felt like I was doing bullshit work. That is hardly a recipe for fulfillment.

My story might not seem important to you. If you are not attending an elite college, why should you care about the social dynamics or the pointless homework? Because this is a story about failing to adapt. I did well when adapting from civilian life to the Corps. I embraced boot camp and infantry school, fully immersing myself. Going from the Corps to school, on the other hand, was much more challenging. I was stuck in the past, not willing to do the work to adapt to this strange new world with its own distinct language, rules, and behaviors. I was a Marine at a school rather than a student.

ADAPT TO THRIVE

Adapting to school is easy compared to adapting to startups. Startups are an entirely new world. A startup has its own culture that is distinct from either the military or college. There are ways of thinking, speaking, and behaving that stamp you as "one of us" versus an "other." If you can't sufficiently adapt to these cultural practices, you will struggle. You will not be able to build and deepen relationships with colleagues, which will undermine your performance and your team's performance, in turn undermining your career trajectory.

Fortunately, you have already overcome a similar challenge in the military. You set an entirely new baseline, discarding a lot of your flabby civilian habits to remake yourself as a soldier, sailor, airman, or Marine. Now, you will do it again. That process begins with taking stock of what you have from your time in the military. Keep what's valuable, as we discussed in the last chapter, and forget what's not valuable. You can then start to add new knowledge and skills to complete the transformation into a Startup NCO.

At this point, you may be tempted to look for a list of the various things you should forget. We actually will discuss a few specific examples to help you stimulate your own thinking on the topic. But my list isn't the same as your list. I had to work—and am still working—very hard to forget specific things. My issues are based on me. They have to do with my experiences and my reactions to those experiences. Your issues will similarly be based on your past experiences and reactions. The list that follows is meant to get your brain working; it is not a complete set of unhelpful behaviors.

FORGET THE DARK HUMOR

As my story about the sexual harassment training illustrated, enlisted folks tend to develop thick skin. We get comfortable making offensive jokes and being incredibly rude to the people around us. This works—for the most part—because we all share a deep bond from serving together. We don't realize how we appear to others, though. That can become a serious issue when we try to work in small teams that include civilians. We need to go out of our way to build strong relationships *before* we start making those bad jokes. I have lost friends because of offhand comments.

FORGET "THANK YOU FOR YOUR SERVICE"

This phrase has been repeated so many times that it has lost all meaning. Of course, I'd rather be bombarded by gratitude than spat on like the guys returning home from Vietnam. We should be happy that the majority of citizens support the military and its mission. But that vague sense of gratitude will not get us anywhere in life. We can't expect that armchair patriots will help us network our way to a great job.

I learned this the hard way when I met people who promised to help but never responded to my follow-up messages or phone calls. No one is responsible for our successful transition but us. Others can nudge and help at times, but their appreciation for our service will not replace our own hard work.

FORGET YOUR RANK

There is no point in classifying ourselves or others as officers or enlisted anymore. This kind of thinking will only hold us back. You can certainly appreciate the value of people's experiences, but do not let it infect your thinking as a civilian. I have had to hire, manage, and even fire senior officers. I would not be an effective leader if I kept any sort of lingering sense of inferiority due to my status as an enlisted veteran. Of course, this mindset shift is not easy. None of these changes are easy. Dropping rank is something I wrestle with to this day, even after years in the private sector.

FORGET YOUR MOS

We cannot think about ourselves in terms of the job we had in the military. That job does not matter anymore. It quickly becomes a way for others to narrowly pigeonhole us into specific civilian roles. The most obvious example from the Iraq and Afghanistan era is former grunts who were guided into private-security or shooter jobs. If you fall for that trap, it will take years to get out. You are not your MOS.

Instead, we must learn to appreciate the skills we developed while in the military. We must learn to talk about those skills, directing people's attention away from stereotypical military roles and toward more productive opportunities. For example, during early job interviews, I emphasized the negotiation skills I'd honed during my Civil Affairs deployment. This always got the other person nodding along about the importance of negotiating a deal so both sides win. The other person and I could build from there, now sharing a sense of commonality.

FORGET YOUR CAREER PATH

The military teaches us to think in terms of linear career progression. You do one thing for a few years, you move on to the next step up for a few years, you go to some education program, and so on. This kind of conveyor belt is not how things work in a modern business, let alone in the startup community. Most people are not qualified for their roles in early-stage startups. Everyone is running around trying to get things done, wearing many hats, and learning on the fly. (*Reminder*: This is one of the reasons why early-stage startups are such a great fit for NCOs!)

Our goal as Startup NCOs is to throw ourselves into the center of the biggest mess and sort it out. That is how we learn as much as we can while building a reputation for effective problem-solving. The first all-nighter I ever pulled was at my second startup. I bonded with my colleagues while showing them I was a hard worker.

FORGET YOUR SUPPORT STRUCTURE

It's impossible to appreciate how good we have it in the military until we leave. Yes, there are a lot of issues, but consider the benefits: we receive a steady paycheck, complete job security, great benefits, lots of perks, free housing, discounted meals and entertainment, and a whole lot more. All this disappears when we take off the uniform, and that comes as a shock to most veterans.

In fact, many people wisely choose college right after leaving the military in order to reproduce much of that support structure. I needed a lot of help from my parents and siblings to figure out how to deal with the random parts of life that used to be done for me—everything from doing taxes and selecting health insurance to figuring out how to sign a lease on an apartment.

FORGET ABOUT LEADERSHIP

Civilian bosses are usually not like leaders in the military. They tend to be much less concerned about personal matters and rarely feel the need to put the welfare of subordinates ahead of their own. They will probably let us down in a variety of subtle ways, especially when it comes to life at a startup. Managers in that setting tend to be less experienced, and they have fewer processes to guide them.

Of course, there are exceptions, but it is important to manage expectations. A civilian boss is not a leader. I already mentioned that the investor in my second startup yanked the funding and didn't pay us for weeks. He was a jerk, but fortunately, my boss was a stand-up guy and refused to accept his check unless the rest of us received back pay. To be clear, that strategy did not work. I am still waiting on those paychecks, but the gesture meant a lot.

READY, SET, FORGET

We often hear that we need to learn in order to grow, but we don't hear anything about how we also need to *forget* in order to grow. You need to make room for the new stuff. Unfortunately, the process of forgetting is painful. It is also necessary. Forgetting is a key step that you cannot skip if you want to become a Startup NCO. You have to dig around inside yourself to figure out what's there and whether it is useful or harmful. There is no shortcut for this process of self-evaluation. And no one else can do it for you.

Everyone has baggage, and Startup NCOs are no different. The military is an intense environment that leaves us with some predictable "scar tissue" based on common experiences. Fortunately, most of these lingering impacts do not affect us in any serious way. They actually have the opposite effect: instead of weighing us down, these experiences build us up. We are stronger, more capable, more resilient, and more confident because of what we've been through. Those are all important traits for a Startup NCO. The tricky part is identifying places where our scar tissue leads to problem behaviors. In those cases, we need to start down the long, hard road to forgetting.

Forgetting takes more courage and stamina than any one person can muster. We cannot go on this journey alone. Forgetting will require the support of people who love us. That does not necessarily mean a lot of people; I've seen examples of people who get everything they need from one other person, usually a spouse or sibling. I've also seen people who spread their support among dozens of people from church, sports teams, neighbors, and so on. Most of us fall somewhere in the middle. What matters is not the number of people but the effect they have. We need to get out of our own heads and talk through our military experience with others who care about what happens to us, confronting the worst parts of ourselves and committing to grinding away at those parts. For how long? Until they stop undermining our lives.

Maybe you raise your voice when you're challenged in a group setting. Maybe you make hilarious—but disgusting—jokes when you're nervous. Maybe you disparage yourself when complimented. Maybe you don't feel comfortable asking questions of senior folks, especially officers. If any of these traits sound familiar, you are not alone. I dealt with all of them myself and continue to fall back on these bad behaviors in stressful moments.

For you, the rules are about to change. Some behaviors that are tolerated—and even rewarded—in the military can get in your way when you join a startup. It will take time to forget them, but doing so is worth it. We need to keep changing, evolving, and improving. We are never done. There is always more to forget so that we can make more room to learn. That is the subject of the next chapter.

SEVEN

LEARNING

When I shipped off to boot camp, I didn't know how to march. I also didn't know how to shoot. The more time I spent in the Marines, the more obvious my ignorance became. There were a whole bunch of things I didn't know. The sheer range was stunning. It included rank structure, land navigation, crew-served weapons, room clearing, darning socks, and cutting hair.

It was not just military stuff either. I learned from my roommate, a Black guy from South Carolina, that I did not know anything about stand-up comedy. (*Side note*: If you don't know about Bruce Bruce, I feel sorry for you!) I also did not have various random life skills: registering an out-of-state car, gathering medical documentation, and getting out of a lease. These were things I had to pick up on the fly as life demanded that I get my act together. This is not just me, obviously—all teenagers and twenty-somethings are constantly dealing with that steep learning curve. It's part of growing up.

Joining a startup takes this dynamic and increases it tenfold. Startup NCOs need to push through an astonishing range of situations, one after the other, with no real pattern or predictability. The only constant is the pressure to perform without having all the knowledge you need. This pressure comes from multiple angles, simultaneously and constantly. Startup NCOs choose this environment. It's very hard and doesn't get easier over time—we just get better.

I'll talk more later about the types of challenges that naturally emerge when working at a startup. For now, it is enough to focus on the essential similarity between startups and the military: the demanding, chaotic, specialized environment. We learned to thrive in the military, and we can do it again. The details are different, but the basics are the same.

LEARNING CURVES

I was still a student when I first started figuring out how to work at a startup in Silicon Valley around 2011 to 2012. There were startups all over the place working on the latest fads, from social media to the sharing economy. Thanks to introductions from some other veterans, I was offered opportunities to work at a couple of different companies. These were not glamorous or highly paid. I was an independent contractor, which means (a) I had no health insurance and (b) I could be fired at any time. I did not understand any of that at the time, of course, but it worked out okay. The upside of being an independent contractor is the ability to legally work for more than one company at a time. That arrangement was helpful to me because I was still shopping around to find a startup job that was a great fit.

No one actually asked me if I was a contractor. They just assumed that I was, and someone from the human resources department at that company sent me a form to fill out. I googled the name of the form, read a few articles about it, talked through a few questions with my brother, and then filled it out and sent it back. That was it. Later, I realized that I could have discussed the terms. I might have been able to negotiate a higher hourly rate. None of that mattered much in the end. Looking back, the money I was being offered is a joke compared to what I make now. The point was learning. I had to get started somewhere.

Part of being a Startup NCO is making mistakes the first time. When it came time to start my first company, I was the youngest guy on the founding team. I deferred to everyone else as we talked through the details. We had drawn-out conversations covering dozens of random issues, but the things that really mattered—equity stakes and the operating agreement—were barely discussed. I did not know where to focus, but I listened, absorbed, and learned. Once again, I had to google a bunch of unfamiliar words, draft a list of questions for people I trusted, get comfortable with the essentials, and then plow ahead with the assumption that no one was screwing me over. More than ten years later, I do not regret those decisions. I put my faith in people who were worthy of it.

For another company I started a few years later, I had to handle all the paperwork. This was a real headache. I had a larger network and some basic business experience at this point, but I still didn't understand the nuts and bolts of creating a business. There are dozens of decisions that I faced for the first time. Different legal structures have different tax implications. I chose wrong the first time. Eventually I had to reincorporate that business, which meant reregistering it with the correct legal structure, changing the state where it was incorporated, paying a bunch of extra taxes, and then complicating my personal tax returns. It took three years to fix everything that resulted from that one stupid decision.

READY, FIRE, AIM

These are just a few examples of what life is like as a Startup NCO. I was—and still am—unqualified to handle one or more situations every day. I needed to quickly get up to speed on some random topic, build or leverage relationships to challenge my thinking, and then figure out how to take meaningful action. This basic process continues to this day and does not show any signs of changing or slowing down. The problems get harder and the solutions get more complex, but I am still just putting one foot in front of the other.

At its foundation, startup life is about figuring things out as you go along: ready, fire, aim. This approach aligns perfectly with the "on the job" mentality that NCOs also have to employ. Think back to MOS training: How much valuable information did we learn in a schoolhouse? Very little, if anything. Most of the important skills we learned as NCOs came from applying basic knowledge. The secret is not classroom knowledge; it's practical application. Startup NCOs need ruthless discipline to grind it out until they've acquired enough experience to get the job done in a more efficient way.

Every startup will challenge you. You cannot predict the specifics ahead of time. All you can say for sure is that you will be thrown into the deep end. You will have a million questions. You will not have enough time. You will have to learn from other people without being told how. And you will have to quickly shift from being reactive to being strategic. If this sounds like an intense but fun work environment, then you are definitely going to enjoy life as a Startup NCO. The grind is all day, every day—and worth it.

Learning quickly is the flip side of acknowledging ignorance. If you know that you don't know, then you have freed yourself to learn. Many people struggle with this self-awareness and humility, especially highly educated and credentialed folks. Many officers, for example, fit well into corporate jobs with a fancy title and a high salary, but they struggle in the early stages of building a business. This is the key area where NCOs have a massive advantage over everyone else. We do not get paralyzed by analysis. We do not overcomplicate everything with fancy theories. Instead, we simplify the work and then rally the team to get it done, setting the example by pushing harder than everyone else.

When you're trying to grind it out at a startup, no one cares where you went to school. Fancy diplomas with

Latin mottoes do not make money for a company. No one cares about PowerPoint slides or fancy spreadsheets with long-term revenue calculations. All that stuff is effectively useless during the chaotic early years of a fast-growing business. What matters is the team figuring out how to get results, which is precisely where the strengths of a Startup NCO shine through. You must learn to repurpose your current skills and become an unstoppable learning machine.

LEARN TO TAKE NOTES

You'll know nothing when you begin at your first startup. Your rate of learning makes all the difference in the crucial early months. The single most effective way to learn is to take good notes. Take the time to capture what is happening in each meeting throughout the day. This will set you up for success, no matter your job.

Take notes in a physical notebook, not on a laptop. Many studies have shown that you'll remember your notes better that way.[15] This habit also helps you in two other ways: First, it shows other people that you are actively listening to everything they have to say. Second,

---

15 Aya S. Ihara, et al., "Advantage of Handwriting over Typing on Learning Words: Evidence from an N400 Event-Related Potential Index," *Frontiers in Human Neuroscience* 15 (2021): doi.org/10.3389/fnhum.2021.679191.

it avoids the temptation of screwing around on the internet instead of paying attention to whatever is going on right in front of you.

LEARN TO FOLLOW UP

Now that you have notes, learn to do something with them. Take decisive action as soon as you can. You will quickly set yourself apart from others if you are seen as someone who is reliable and fast. The most impressive and valuable reputation you can have in the startup community is that you get shit done.

Make a note of everything you were told to do and everything you told others you were going to do. If these tasks can be completed in a minute or two, make sure to do them *immediately*. Use a different approach for longer, more complex tasks: schedule a recurring thirty- to forty-five-minute block of time in the morning of each workday to complete all remaining tasks. This basic time-management skill is simple and effective, yet almost no one does it. Be the one who does.

LEARN TO SELL YOURSELF

Every branch of the military is rooted in warrior culture. We do not like or respect people who talk about themselves in a positive light; we want others to recognize our contributions. There is a simple way to confirm this: Imagine someone putting themselves in for a medal. How would you view that person? As someone rightfully claiming recognition for high performance, or as a self-absorbed jackass?

In the military, we have leaders who are expected to ensure credit for members of their teams. There is no parallel in the startup world. No one else is responsible for our career, so we need to advocate for ourselves. That means understanding the value we create in a work environment and explaining it to the people around us. Basically, it entails educating others about what you can do for them. Think of it like the instruction manual that comes along with a product you buy. Otherwise, how will anyone know what you can do?

LEARN TO HELP PEOPLE

There is a powerful way to ensure you have a successful career: take the time to help the people around you. Some businesses say they have a "no asshole" rule. That is a great start, but it does not go far enough. Force yourself to reach out and help others. Do this for years, and you create a halo around yourself. If the first startup you join fails, and the second startup you join fails, and even the *third* startup you join fails, you will be fine. You will have multiple offers before you even realize you need to look for a job. Your halo will protect you.

Helping others is simple. Practice the same ritual to end every meeting you have: Ask the other people there, "What can I do to help you?" Listen and follow up with questions to make sure you understand. Once people see that you truly want to help and actually know *how* to help, they will open up to you. Over time, you'll build up a massive amount of goodwill. Everyone will like you and want to make sure you are taken care of, no matter what.

LEARN BUSINESS JARGON

The military is full of annoying acronyms. So is every industry, every department, and every company. Every group of humans takes great pleasure in adding highly specialized words that have no meaning to normal people. Expect this and embrace it. The faster you learn these words, the faster you will fit in with the startup community.

When preparing for a meeting, review the agenda and look up any words you don't understand. Look up and memorize all the new words that you hear in meetings. If you hear a new word on a Monday, then by Tuesday you should be using that exact same word as if you'd known it your entire life. This goes a long way toward establishing you as a legitimate member of the team.

LEARN HOW THE BUSINESS FITS TOGETHER

Each business is unique in its own way, but they all share certain characteristics. There are people who pay money to the business; these are customers. There are people who create and deliver things for those customers; these are engineers. There are people who try to convince more people that they need the thing you make; these are salespeople. Each group has a different culture, different expectations, and different patterns of life. Some roles reward action, while some are more deliberate. Some roles require a lot of travel, while others require workers to be in the office every day. Some roles allow for more individual work, while others are primarily in teams.

Do not get pigeonholed when you start working. Make sure you meet people from all parts of the company. Talk to them about their lives. Learn what their schedules and meetings are like. Learn how they are compensated, what they did before this job, and why they took this job. Learn what they want to do next. And, of course, learn how you can help them (see above).

LEARN ABOUT THE MONEY

Businesses exist to make money. At the end of the day, a business is a legal container for money to flow in from customers and flow out to employees and shareholders. Do not forget this basic fact as you begin your journey into the business world. You need to understand how the money moves around. Very few employees truly understand the flow of money inside their own company.

Take the time to research companies like yours. Read about the large public companies that operate in the same industry. Listen to podcast episodes about the industry. Read general business books and industry-specific business books. Craft a list of questions about the flow of money through the organization, from the customers who send it to the suppliers who receive it. How long does the cycle take? Who is involved? How does your piece of the puzzle fit in?

LEARN TO ENGAGE THE TOP

NCOs should not be known by the command officer of a battalion, regiment, or ship. If they are, it's usually because they screwed up. Mostly, NCOs try to fly under the radar and focus on getting the job done. At a startup, you need to think differently. There is a never-ending stream of opportunities flowing through the company. To seize them, make sure you build relationships with the founders and senior executives. Doing so will help you plug into those opportunities and, if successful, receive credit for great work.

Do not let anyone else control your relationship with the top. It's human nature to take credit for the good and avoid blame for the bad. Your boss may be a good person, but that doesn't mean they are an angel. Only you can ensure the people at the top clearly see and understand your contributions. Reach out to them on LinkedIn. This is not aggressive behavior but rather "Wow, that person is hungry!" behavior. There is a difference, which I'll cover in Part III.

LEARN FAST

If there is one thing that people need to do at a startup, it's to learn fast as a group to discover the right customers, what they need, and how you can deliver it to them at a profit. Anything else is a distraction: strategy sessions, designing logos or websites, building product roadmaps, et cetera. That stuff may eventually matter, but it doesn't at the beginning.

As a Startup NCO, you will need to set the pace for this rapid learning. The CEO won't hire you or promote you because of your qualifications. They will only do that if they know they can rely on you to figure out what needs to get done and then go do it. You would be surprised how hard it is to find people like that. Degrees from places like Harvard and Stanford will help people get an interview, of course. But the degrees do not help them adapt on the fly when the company needs to rebuild itself or pivot to a new business model.

At first glance, a Startup NCO may not seem all that impressive. Our clothing won't be the same. We'll be rough around the edges. In particular, our speech and behavior may rub some people the wrong way. But over time, a Startup NCO will leave everyone else in the dust. There is a certain relentlessness that other people admire but struggle to replicate. We never stop learning. We focus on the fundamentals, pushing them to the maximum.

After my initial enlistment, I knew I wanted to go back to school. I had flunked out at eighteen years old before joining the Marines. Now I thought maybe I had built up the skills I needed to do okay in the classroom. I was still scared, though. I worried about competing with other kids who had done well in high school, so I tested the water. I went back to a community college for a summer, only taking two remedial classes. I went to every office hour with the professors. I worked twice as hard as anyone else in those classes. Despite all that work, I still saw myself as a dumb kid who had flunked out of school. That sense of inadequacy pushed me to try even harder. My day of reckoning came at the end of the summer. I had to recheck my grades to make sure the teachers didn't make a mistake: two A's.

When I graduated from community college, I transferred to Stanford University. It was my top choice out of the five universities where I'd been accepted. I had also maxed out the number of courses I could take each semester, earned straight A's, and finished as the top student in my entire college. But I was still the same person who had sweated those remedial classes less than two years earlier. I was not the smartest student there and didn't need to be. I had adapted the discipline from the military and applied it to school. Even back then, I sensed that this was a repeatable process. I could make this sort of adaptation again and again. If I could figure out how to learn faster than anyone else, I could always bet on myself. The steeper the climb, the faster I'd rise.

Everything in this book has laid the groundwork for what comes next. You now know more than almost any other NCO before you as they prepared to leave the military. You have a broader perspective on how to build a new life for yourself after the military, based on the experiences of many NCOs who have successfully done this before you. Their hard-earned wisdom is yours.

Now, it's time to move to the part that should be the most natural for us, as well as the scariest. Now, it is time to act. You must go through a challenging journey, the Startup Crucible, that prepares you for the next season of your life. The rest of the book will walk you through that process, offering step-by-step guidance on weaving together your military experience with your personal goals to build a supportive network and find a job at a great startup.

PART III

THE STARTUP CRUCIBLE

When we enlisted in the military, we signed up to be a part of something bigger than ourselves. We saw a bright future ahead, and we were willing to work for it. We knew we would have to earn our spot, not have it handed to us.

That's when we were shipped off to boot camp for weeks of intense training. The goal was to break us down individually before building us back up as a part of a successful team. And we each made it through that crucible, even if it sucked sometimes.

It takes over a hundred days to get through basic training. More advanced levels of training can take months, even years. There are clear standards we need to reach for our MOS, billets, and special skills. No matter what we do during our tenure in the military, our peers judge us according to those standards. We must be proficient in everything from military knowledge to physical fitness and from ethical conduct to combat lethality. At each stage, there are milestones that let us know we are on track. None of this work is easy, but the military sure does make it simple.

There is no equivalent to boot camp or advanced training to transition out of the military. The various branches of service put a lot of thought into how they train people to put on the uniform. Unfortunately, the process of taking off that same uniform is an afterthought. We get a few PowerPoint classes, some generic advice, and a bunch of information that we will likely never use, and then they show us the door.

When we first transition out, we are not ready to work at most civilian jobs. We lack the qualifications and experience to walk into a demanding environment on day one and create value. Interesting jobs that pay well are not easy to get, precisely because so many people want them. It takes time and effort to compete and win those job opportunities. The lengthy adaptation process helps us reshape our NCO skills and experiences to fit the business world. This process is especially critical if you want to work in startups, which face the additional demands of rapid growth and internal chaos.

You might consider this bad news, but it's not. Far too many people leave the military thinking they are ready for a great job. In fact, they think they *deserve* a great job. They have too much "Thank you for your service" ringing in their ears. That sort of entitlement ruins your work ethic and actually makes you a liability. Do not be that person, or you'll make the rest of us look like entitled fools.

If you're struggling to understand this concept, try flipping around the situation. Imagine there is an incredible entrepreneur who has started several companies that are all worth millions or even billions of dollars. This person is widely celebrated in the press, respected by peers, and loved by employees. Now, imagine that person being shoved into command of a military unit, particularly one headed into combat. How well do you think they would fare?

Obviously, they would do a terrible job. They would feel overwhelmed, and the staff would have to take over or risk the lives of the people in the unit. That lack of success wouldn't be based on any inherent failing on the part of the person. Recognizing a poor fit isn't about assigning blame; it's about acknowledging the difficulty of transitioning between environments. Adapting skills and experiences to a new context takes time, no matter how talented or hardworking you are.

Now for the good news. You are approximately six months away from getting a great job at an awesome startup. That is about how long it takes to make it through the Startup Crucible. You can emerge on the other side as a Startup NCO. To get there, you will take the work ethic, skills, and values you picked up as an NCO and strategically adapt them to this crazy, fascinating, and wonderful world of startups.

FRAMING THE CHALLENGE

At this point, the concept of a Startup NCO probably seems pretty abstract. What do these words mean in practical terms? It is actually quite easy to judge whether someone has become a Startup NCO. The definition is a functional one. The only question we need to ask is "Are you performing well in a great job at a great startup, using your military skills and experiences?" If the answer to that question is "Yes!" then congratulations, my friend, you are a Startup NCO. If not, we still have some work to do.

That work starts by breaking down the three key parts of this question:

1. Think about the great job.
2. Think about doing well at the job using your military skills and experiences.
3. Think about the great startup.

We will begin with the first part: the job. You need to have the job! That means someone has offered you a job and you have accepted the offer. That is the part where all of us tend to focus, but all three aspects of the question are important.

There are lots of ways to get jobs, most of which rely on factors outside your control. Some folks get lucky as they leave the military—they trip and fall, landing on an awesome job. Some people happen to have great connections through family members or friends. Others are water walkers who are so talented and gifted that recruiters at great companies are already excited about hiring them.

Most of us cannot rely on any of these options. I certainly couldn't when I first started out. Jobs for people like you and me will not fall from the sky. We need a plan that we can follow, one that has actually been proven to work in the real world.

At the same time, it is crucial not to skip over the second part of the question: doing well at the job you have. You aren't set up for success if you trick someone into giving you a job and then flail around and get fired. That is even worse than not getting a job at all. Reputation is everything in the startup community. A Startup NCO must be able to deliver.

THE RACE TO 250 CONVERSATIONS

So, what do we need? A plan to go from point A (where we are today with our military mindset and skills) to point B (where we will be in the future with a Startup NCO's mindset and skills). The Startup Crucible is that plan to turn you into a Startup NCO. It offers you a concrete path with specific actions you can take every day that build from week to week. The Crucible's goal is for you to thrive in a job at a great startup. Like military training, it will be easy to read about the steps you need to follow but hard to implement them in practice. That's why I'll discuss each stage in detail, breaking everything down into bite-sized chunks.

Here is your introduction to startup thinking. This is called "napkin math," meaning it is a simple but useful calculation that can help you cut through a bunch of noise. Napkin math is so simple that you can do it on a napkin. It is also so simple that an enlisted Marine like me can do it! Here's the napkin math for the Startup Crucible: 250 conversations × 2% conversion = 5 job offers in 6 months.

250 conversations × 2% conversion =
5 job offers in 6 months

To get you to that great startup job, you need people to offer you jobs. You won't say yes to just any option, though. You want to have at least five job offers that excite you. Those offers will come after a lot of hard work. You will have to build dozens of high-quality relationships with people who appreciate your skills and want you to land a great role.

Unfortunately, you probably do not know any of those people right now. That means you can plan to have at least 250 conversations to find and convince them that you're the real deal. Why 250? There is a 2 percent conversion rate from initial conversation to job offer. This is based on a combination of my personal experience, independent research, and conversations with other Startup NCOs. And yes, this will all happen within the next six months.

The odds are currently not in your favor. There is no reason to sugarcoat things for you. As an NCO, you know what it takes to get stuff done. The world is a busy place, and no one is sitting around thinking about how to help you, except maybe your mom. You are a nontraditional candidate for any job you pursue. That is another way of saying that you probably won't get the job even if you apply. So, you need to outthink and outwork others. Your creativity and determination will carry the day. Everything I'm suggesting is possible if you put in the work.

Don't worry if this seems like too much. It's actually very doable if you follow the steps laid out in the next few chapters. Right now, all you need to do is imagine how it will feel to have that great job. It is a truly wonderful place to be. Imagine all the best parts of the military—the mission and sense of purpose, working with people you like—with all the bullshit removed. You will have the confidence that comes from successfully making the transition out of the military. No one can take that away from you. You will always have options because you know, deep down, that you can build them for yourself. You will be unstoppable.

Six months of hard work lie ahead of you. Your Startup Crucible awaits. Now is the time to commit. There are four major phases of the Startup Crucible, which I'll go over in the next four chapters:

1. **Exploring**, where you get to know different industries and companies—you're on the hunt for teams where you could become a true missionary and be on fire for the work every day.

2. **Cultivating**, where you start having in-depth conversations with key leaders at startups that interest you—you'll learn how to give and get something out of these conversations so that you don't worry about reaching out to awesome people.

3. **Deciding**, where you evaluate the options that you've developed—this is a difficult phase that requires some critical thinking and blunt conversations.

4. **Onboarding**, where you immediately prove your value to your new startup—you're setting yourself up for long-term success with your boss and peers.

Remember the goal: doing well at a great startup using your military skills and experience, not just getting a job. This is where you bring everything together. If you hit that target, it proves that you've built a professional network and adapted your NCO skills to the startup world. You will become a Startup NCO, and nothing can stand in your way.

EIGHT

EXPLORING

If you want to make a lot of money doing interesting work, remember one simple truth: jobs come from people. It's amazing how many people forget this basic fact or act like it isn't true. Jobs do not float around in the air—or on the internet. They do not exist in the physical world like a car or a phone. Jobs exist inside businesses, and businesses are run by people. If you want a job, you have to get to the people at the business where that job exists.

Why is this so important to keep in mind? Because now you should be crystal clear on the main activity that will help you find great jobs. What is it? What single activity is the most critical? Talking to people who work at businesses that have jobs you want.

Stare at this statement and think about it: talk to people who work at businesses that have jobs you want. This is a big idea with serious implications for how you spend your time while job hunting. Most people who look for a job are wasting their time. They tend to focus on surface-level activities such as writing résumés and applying to jobs using online platforms. That takes many hours and will not get you any closer to the critical conversations with people who can offer you a job that you want. A job application is something you fill out *after* that conversation, not before.

Many people struggle to understand this fact, so we will spend a fair amount of time on it. To be successful, spend your time on productive activities instead of wasting time on unproductive ones. When it comes to getting a great job, the single most productive activity is talking to people. Lots and lots and lots of people. Why? Because you will meet people who offer you a job. How does that happen? People offer jobs to other people they like. And how does one person end up liking another person? By talking to them, enjoying the conversation, and being impressed by them.

Here's another way of explaining the dynamics of getting a great job: no one is going to review your job application and then decide to hire you. That simplistic model misses steps, and it also gets the order wrong. First, the person will figure out if they like you. Then, if they like you, they will decide whether they can justify the decision to hire you. Finally, you will figure out how to capture your skills and experience on a résumé that provides the evidence to support the decision.

Emotion → Decision → Justification

Do not mix up this order. First comes the emotion. Then comes the decision. Then comes the justification. This order feeds directly into an important set of questions:

1. How do I find people to talk to?
2. How do I get them to like me?
3. How do I help them justify the decision to hire me?

I'll cover the first question in this chapter and the second and third questions in the next chapter.

THE MAGIC NUMBER

During the Exploring phase, the focus is on quantity, not quality. You want to have lots of conversations rather than just a few. You do not need to be highly selective at this point. How many conversations are we talking about? At least 250 of them. That sounds like a lot, but if you break it down to an average number per workday, it's only two. Assuming five working days, that translates to ten per week. You will hit 250 in twenty-five weeks, which is less than six months. Two conversations per day take up about an hour. Will you invest an hour a day to ensure that you have a great life after the military?

> 2 conversations/day × 5 days/week × 25 weeks = 250 conversations

You want to explore and to learn during this phase, not take the first job you're offered. There is a very simple reason for this. The first offer you receive is almost certainly not a good fit for you. You do not want to rush into the wrong job. Chances are, you are aiming too low. There are better jobs out there if you take additional time to discover what the labor market has to offer. Listen to lots of people tell you about their lives, their career milestones, their major decisions, and other valuable insights. The more you hear, the more you will pick up on patterns about the work that appeals to you and which choices will put you on the path to get that type of job.

Some people start the transition process with some idea of the kind of job they want, or at least the type of industry. If this is you, stop now. Avoid this temptation at all costs. Do not let yourself get pigeonholed at the beginning of your transition. You are just beginning to learn what's out there. Cast a wide net, talking to people from a variety of industries who work at different levels of organizations. Over time, you can narrow down to earlier-stage startups, but first you need a basic "map" of the labor market. At this point, you probably know only 5 percent of the kinds of jobs that are even available.

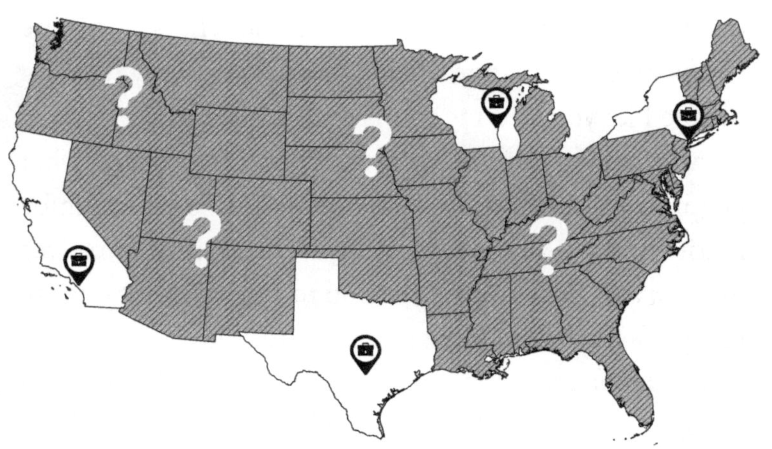

THE LABOR MARKET

A little more context is helpful before jumping into the search for a great job. Our military experience will mislead us if we aren't careful. People working in businesses do not fit into a neatly ordered organizational chart like they do in the military. You cannot know as much about a person based on their job title. The policies and regulations that create so much predictability—and rigidity—in the military do not exist in the private sector. Rank and MOS can tell you a lot about what someone in the military has done and how they got to that point.

Instead of a clear military hierarchy, the civilian world has a confusing set of overlapping networks called the labor market. People are constantly moving around from company to company. Even the people who stay in one company often change roles. This makes it very hard for people like you and me to make headway when looking for an amazing job. We have to get good at guessing what people do and what they care about. This is critical information for us. We use these insights to describe ourselves in ways that get the other person to care, making it more likely they will respond when we reach out.

The labor market does not make it easy, though. Job titles vary from industry to industry, and even from company to company within a single industry. Some companies are organized in more traditional structures with strict roles and responsibilities, while others are very flat with greater empowerment at the individual level. Different companies will use the same titles—such as "product manager," "account executive," or "vice president"—but they will mean different things. The work, skills, mindset, and background will be different. The approach we need to take will therefore also be different. The following page shows one example of a product manager to illustrate the point.

If this situation confuses you, you are not alone. As you begin your Startup Crucible, you will navigate a complex landscape. But don't worry; there's a playbook to follow. There are steps to take. Success is not magic. It just takes creativity, patience, and hard work. As an NCO, you have already proven you can creatively apply basic principles in challenging environments.

GETTING STARTED

There are several easy ways to get started, and they all begin with LinkedIn. This community should be your go-to tool during the Startup Crucible. This is the main platform for professionals. It's amazing how many people you can reach and how much information you can learn through LinkedIn. Signing up takes about two minutes,

PM in Early-Stage Startup

John Smith
Product Manager at My Startup
Los Angeles, California, USA • Contact Info
500+ connections
in

This person works in a small team with almost no process or structure. They will be working 70–100 hour weeks, grinding it out with a few people they probably knew from college or a previous startup.

PM in Growth-Stage Startup

Jane Smith
Product Manager at Mid-Sized Business
Denver, Colorado, USA • Contact Info
500+ connections
in

This person works across 2–3 teams. There is some process and structure. They will be working 50–70 hour weeks. They will work with lots of stakeholders and use advanced software tools to do the work.

PM in Late-Stage Startup of Big Tech Company

This person works across many teams. They have to work within rigid processes and structures. They will be working 40–50 hour weeks. They will report to multiple bosses and use more traditional tools to do the work.

which you can do by going to *www.linkedin.com* and clicking on the "Join Now" button. Use your Gmail address to log in. If you don't have a Gmail address, go create one at www.gmail.com. Click on the "Create an Account" button and follow the prompts. Use your first name and last name plus a few numbers for your personal account. Do not create an account with references to the military or your MOS. If you already have an account like that, create a new one. You are trying to adapt to the civilian world right now, not reinforce your military identity. Civilians would not want to email me at *marinekillbodies@emailserver.com.*

Once you create a LinkedIn profile, here's what to do:

1. Upload a picture. Do not use a military picture. Use your phone in portrait mode and have a friend take a zoomed-in picture of you in a collared shirt, smiling at the camera.

2. Add your military and educational experience. Use plain English when completing the initial profile sections. You will have time later to edit and revise based on what you learn in conversations.

3. Include some basic demographic information, such as where you live.

As you create your profile, take the opportunity to find some people you already know who are on LinkedIn. The more connections you have, the better. It's helpful to have at least a hundred connections, otherwise you will seem like a spam account. Even twenty or thirty is enough to get started. Don't worry; you will be adding more connections throughout this phase. Start with people from your hometown, your high school, and your unit. You might be surprised to see some of them are already doing well professionally. These will be good people to target for early conversations.

Now, it's time to start finding new people. These are the ones who will eventually lead you to others who have jobs. LinkedIn provides you with an excellent overview and search function for the American labor market.[16] You can also use these exploring filters: desired location, personal and professional interests, startup trends, recently transitioned connections, and friends and family. Each of them will help you find people for an exploratory conversation. There are hundreds of millions of people on LinkedIn. You are trying to get in touch with a tiny percentage of them. Filters are your friend.

16 For more information, visit https://www.linkedin.com/help/linkedin/answer/a6213379.

FILTER #1: DESIRED LOCATION

Some people already know where they want to end up after the military. The metropolitan region around that place has an established economy and labor market, with certain dominant industries. Examples are aerospace in Los Angeles, enterprise IT in Silicon Valley, finance and media in New York City, and biotechnology in Boston. Startups in the area will often work in that industry since they hope to someday compete with—or be acquired by—the industry giants.

You will also find research universities supplying human capital and companies providing industry-based products and services. These are well-developed communities that combine money, people, and ideas. The result is a constant stream of new startups hoping to become the next big thing.

Here's your first bit of startup jargon: we call these places "ecosystems." Location is important for you to consider because individual startups may not be successful, but the ecosystems endure. You will find many opportunities through the relationships you build in an ecosystem. Alternatively, if you want to return home or remain in the area where you're currently serving, that is an option too. Many startups will be limited to in-person roles, but there are also opportunities to work remotely or in a hybrid environment (meaning some work is remote and some work is in person).

FILTER #2: PERSONAL AND PROFESSIONAL INTERESTS

You might be working on passion projects on the weekends. You might be reading an endless stream of dense technical books. You might be subscribed to subreddits or blogs. You might belong to a Meetup group or watch endless YouTube videos on some obscure topic. If that's the case, then you should at least consider whether your weird obsessions could mature into a job. This can be a scary thought for many people leaving the military. Creativity tends to get beaten out of us because the culture is so focused on getting things done. Your transition is a great time to open back up. Give your artistic side a chance to restore itself. These are muscles you have not used in a long time.

Of course, there may not be a job waiting for you in the field of restoring antique birdhouses or yodeling. That is not the point at this stage of your Startup Crucible. It is worth talking to a few experts in any field. Try to learn whether (and how) they make money. Start with Google. In the search prompt, include your interest and the word "LinkedIn." Google usually does a better job of finding people than if you search the same thing directly on LinkedIn.

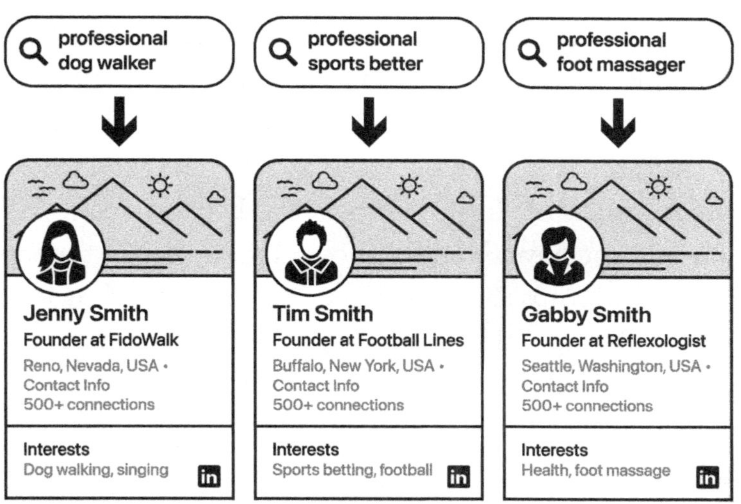

FILTER #3: STARTUP TRENDS

Technology sectors are cyclical. There is a lot of boom and bust. Investor money comes in waves, and so do the great jobs. Looking back over the last few decades, we had the personal computer wave in the eighties, the consumer internet wave in the nineties and aughts, the social media wave of the teens, and so on. There have also been smaller cycles: green/clean technologies, software as a service (SaaS), the sharing economy, augmented reality, virtual reality, blockchain protocols and cryptocurrencies, and artificial intelligence (several times). As I write this, defense technology is actually in the middle of its own cycle.

Some game-changing startups survive these cycles, but most do not. This matters to you for one specific reason: lots of money equals lots of jobs. There are tons of opportunities to "ride the wave" as investors fund a host of companies hoping for the next Facebook or Google. As a result, you have a good chance of joining one of the early-stage startups that have recently received funding and are looking for talent. The founders of these companies will be excited to hear from someone who is attracted to the challenges of a fast-growing business.

FILTER #4: RECENTLY TRANSITIONED CONNECTIONS

There is something especially useful about hearing from people who are a few steps ahead of you. Seek out people from your unit who have left the military in the last few years. Even someone who got out six months ago can be worth a conversation. These people tend to have more practical advice than those who are farther along in their careers. They are grappling with all the challenges of transition that you will soon face. They know what is—and is not—working in their own transition. Some of these pain points and lessons learned will clear up nagging questions. Other insights will force you to think more deeply about your own goals.

One note of caution: it is easy to listen too much to people like this. Be careful that you do not try to mimic them. Listen to them for insights about how they started to build their networks and learned to translate their skills and experiences. It is okay to dig into the specifics, but remember that they are different people with different goals and values. This is especially true when it comes to officers. Many of them will want to help, but their backgrounds and post-military professional opportunities do not match yours. They will either offer useless advice about getting an MBA or going into investment banking (when you do not have a bachelor's degree) or talk down to you. Learn to listen for patterns, not for advice. Remember, this is the Exploring phase.

FILTER #5: FRIENDS AND FAMILY

There are billions of people in the world, but only a dozen or so who know and love you. It is important to take the time to listen to them, but probably not in the way you imagine. Do not ask them where they think you should work. Instead, talk to them about who you are as a person.

Family and close friends can help you think through your journey—from childhood, to when you went into the military, all the way to today as you are getting out. Ask them to highlight the parts of you that have not changed over the years. This broader view will help you identify the ways you have changed, while also appreciating those qualities that have remained constant. These people will usually do their absolute best to support you. They have their own agendas, though, which might not align with yours. For example, do not be surprised when people in your family advise you to come home rather than explore other areas. There may be fantastic job opportunities in New York City or Austin or San Francisco, but that does not mean anyone in your family wants you to move there.

REACHING OUT

These five filters will surface dozens, even hundreds of people to talk to. Now you need to reach out to them. What do you say to someone when you first reach out? How do you build a network from scratch? Questions like this point to a phenomenon known as "the cold start problem." Fortunately, this problem is not that hard to solve. You need a template LinkedIn message and a template email. You will modify them as appropriate, based on the person's name, background, and how you got their contact info. Do not write a lot.

LinkedIn lets you send a limited number of messages to people. These are called InMail. Regular accounts do not get InMail—only Premium accounts do. Fortunately, you qualify for a promotional subscription! Whether you are a veteran or still on active duty, hit the footnote for two useful links.[17] Signing up will take a few minutes, but it is worth the time. The Premium account will make your profile look more legitimate to anyone who checks you out. Try something like this for the LinkedIn message when you are reaching out directly:

Subject: [*Job Title*] Career Advice

Message: Hi [*Person's First Name*],

I would love to speak with you about your career, especially your experience at [*Current Company*].

I am leaving the military in six months and am learning about different opportunities. It is incredibly helpful to speak with successful professionals.

Please let me know if you are willing to spend 15 to 30 minutes on a call. Thank you in advance for your time.

---

17 For more information on a complimentary Premium Business subscription for *veterans*, visit https://www.linkedin.com/help/linkedin/answer/a518653/linkedin-for-veterans-free-premium-career-subscription-and-eligibility. For more information on a complimentary Premium subscription for *active-duty military members*, visit https://socialimpact.linkedin.com/programs/veterans/activeduty.

Sincerely,
[*Your First Name*]
[*Your Phone Number*]

You can also try to connect with someone directly, which is similar to mutually following on social media. These connection requests will annoy some people, which is okay. Remember, this process is all about connecting with as many relevant people as possible. The main benefit of LinkedIn's Connect feature is that it does not use up your InMail credits. Click on the "Connect" button then click on the "Add a Note" button. You can write a very short message like this:

Hi [*Person's First Name*], I am excited to speak with you about lessons I can apply from your career to my transition out of the military. Would [*day*] at [*time*] work for a 15-minute phone call? —[*Your First Name*] [*Your Phone Number*]

This may seem aggressive, and you should be prepared for a lot of rejections. It is important to be clear with the person so that they understand why you are reaching out. You want to speak with them briefly about your transition out of the military. Some will say yes and some will say no, which is fine. You are building a "funnel" of people, which means a large number of people need to go into the top for you to get what you want out of the bottom. Here's the napkin math: For every ten people who accept your invite, expect five to respond to your follow-up. Of those five, expect one or two to agree to a specific time and share their phone number. That is a normal level of attrition during this sort of outreach.

Instead of InMail, you can also use email. If someone accepts your request, you can look up their contact information on their profile page. Once you have their email address, try sending them something like this:

> *Subject:* Sharing [*Job Title*] Career Advice for 15 Minutes
>
> *Message:* Hi [*Person's First Name*],
>
> My name is [*Your First and Last Name*], and I'd love to talk to you for 15 minutes. Sorry for the unexpected email. I looked up your email address after we connected on LinkedIn.
>
> I'm getting out of the military in a few months and am learning more about different careers. Your background is impressive, and it would be really helpful to talk briefly so I can learn about your career.
>
> Are you available to speak on the phone this week? [*time range*] works best for me, but I can be flexible.

Thank you in advance, and don't worry if now isn't a good time for you.

Sincerely,
[*Your First Name*]
[*Your Phone Number*]

Do not get confused about your objective here. Most people stress out about these types of emails and write way too much. Shorter is better. The goal of the email is to get the person to agree to talk to you, not to share your life story. You want to be interesting—that's all. They'll forget the other details about who you are and why you are reaching out in a few minutes. When you actually speak with them, there will be a chance to remind them of your story.

Reaching out to people you don't know is fantastic practice. Remember, the Startup Crucible is about finding and doing well in your first job. The process of finding the job helps you adapt your existing military skills to work well in a startup environment. This is your first practical application exercise. Can you get people to care enough about you to schedule a call? If so, you will have a huge advantage over others. Startups are always on the hunt for new customers, and this is the way it's done: find people, and then get them interested enough to schedule a meeting.

PUTTING IT ALL TOGETHER

You need two more tools to go along with your outreach:

1. A way to keep track of all the people you identify
2. A way to keep track of what you're learning

A spreadsheet is fine for most people, and a document is fine for the patterns that you notice over the course of your conversations. You can use Google Sheets and Google Docs, unless you are already familiar with more advanced software. Since you have a Gmail account already, these two tools are available as a part of Google's family of products. As any good NCO knows, the tools matter less than the person using them. A Marine with a plastic spork is more dangerous than a schmuck with a pistol (but don't feel the need to go out and test that particular claim).

Preparation is an important part of reaching out to people during the Exploring phase. Ten to fifteen minutes of background research is fine. You do not need to go all out, but you should show that you did some homework before sending someone an email. People respond well to preparation, as it shows that you are serious and thoughtful. This step dramatically increases the chance that the person will respond, the conversation will go well, and you will expand your professional network—all of which will move you closer to your goal of a great job at a great startup.

Research should help you get the most out of your Exploring phase. Once you find someone you think is worth talking to, skim their LinkedIn profile. Look for jobs or skills that you find interesting. Look for any useful information that pops up when you google the person's name. Write down notes of anything that interests you. You can reference these notes for your introductory email, during the conversation, and then again in a follow-up email thanking the person for taking the time to speak with you. A disciplined approach to exploration will immediately set you apart from the crowd.

During the Exploring phase, you will need to adopt one specific tactic that does not come naturally to most people: asking for more introductions. This is a critical skill for a Startup NCO. Practice it as soon and as often as possible.

Asking for more introductions is like doing ball-handling drills in soccer or basketball. It is absolutely essential to function at a high level. All the talent in the world will not help you if you cannot expand your network.

Ryan Hunter
Senior UX Design Manager at Charles Schwab
Keyword: UX Design

1. How did you first become interested in product design?
2. What is the most interesting part of working on digital experiences?
3. What are the best books or other resources to learn more about UX/UI?

...

THE FREEDOM TO EXPLORE

As you plow ahead in these Exploring conversations, you will also start to learn all kinds of interesting things about the private sector. Remember when you first joined the military? You learned new words, especially acronyms, on a daily basis. The same is true of the professional world, except even more so. The world outside the military is much more fragmented, especially when you start to learn about different startups in different industries.

You will hear about a lot of new ideas, tools, metrics, and ways of working. You may start talking about things such as MQLs (marketing qualified leads), MRR (monthly recurring revenue), or CAC (customer acquisition cost).

Remember, the point of this phase is to explore. It is normal and natural for this part of the Crucible to feel scattershot, even random. You will certainly feel like you are wasting your time after a few of the more awkward conversations. Do not worry, and definitely do not change course. Exploring isn't about efficiency. It's about learning, and that takes time. It also takes some grit. You will feel the pressure to have a clear direction and purpose to all these conversations. You will also feel the pressure to fiddle with your résumé, search job sites, or read generic job search advice online. Those are time wasters. They will not help you. What should you be doing instead? Having at least two new conversations per day, every day.

All this work may seem very squishy and poorly defined. That's okay, because this part of the Startup Crucible is designed to be that way. Exploring is not directed toward a specific end point. There is only one goal: expanding your network to hundreds of people who like you and are interested in your career. That sets you up nicely for the next phase: Cultivating.

NINE

CULTIVATING

Startups are made up of people. Every conversation is an opportunity for you to figure out whether you want to join a specific team. The person you talk to is asking that same question from the other side: *Do I want this person to join my team?* Remember these early-stage companies are still figuring out their culture. Each individual who joins an organization will make a big impact on the values that ultimately get baked into everything.

If you find that you do want to join a team, then your aim is for the other person to want you to join as well. That mutual interest, just a hunch at this point, is the trigger to move forward in the interview process, which will bring you one step closer to your goal during Cultivating: getting a job offer. The more conversations you have, the more you will find interactions with a spark of interest. That's why the Exploring phase emphasizes the quantity of conversations: on a basic level, this is a numbers game.

By the time you hit a few dozen conversations, patterns will start to emerge. You will identify the specific stories you tell that immediately grab the other person's attention. At the same time, you will get stumped by certain questions. That is also normal. You will eventually figure out a good answer to those questions, and they won't stump you anymore. Sticking points are a part of the learning process.

Imagine a hollow-point round as it enters something. It comes in small, then quickly expands. That is what you need to do. Once you establish a single connection at an organization, you should be able to grow your network to accomplish your goal. Right now, your goal is getting high-quality job offers, but eventually that goal will change. Startup NCOs are asked to do everything from assessing cybersecurity risks for a new product that hasn't yet launched to persuading an IT person to approve the implementation of a particular type of software. They achieve success by figuring out how to get to the right person inside an organization, and that starts with a single individual who makes a "warm intro" to the next one, and so on.

EXCITEMENT: THE KEY INGREDIENT

The Cultivating phase runs on excitement—yours and others'. A good Cultivating conversation is therefore one where both you and the other person get very excited. Excitement generates the energy that drives the forward momentum and expanding force that we just discussed. People who are excited about you will overlook issues, work around processes, shorten timelines, and otherwise "hack" their own systems to help you.

Excitement by itself is not enough, though. There are still a lot of details to work through. Energy needs to be harnessed and pointed in a particular direction. After a good conversation with someone, you will immediately start thinking about the next steps. The other person might do the same, but they also have other people at the company to worry about. Your next step is getting those people to be as excited as the first person. How do you do that? With the most important tool that anyone has in the modern economy: email. A good follow-up email can work miracles for you. Here is a template to help you transition from Exploring to Cultivating:

> Thanks so much for taking the time to speak with me, [*First Name*]. I'm especially excited about the next phase of growth for [*Company*]. Per your suggestion, I would love to meet with [*First and Last Name*] to learn more about their vision for the team and how I might fit in.
>
> It sounds like they need folks who can think strategically and roll up their sleeves. I was surprised by how much my experience and current work seem to overlap with those needs. My LinkedIn profile ([*profile link*]) doesn't do it justice. I'll be able to explain more on a follow-up call.
>
> Sincerely,
> [*Your First Name*]
> [*Your Phone Number*]

Do not skip this step. If you do, you are unlikely to become a Startup NCO. In fact, you will likely get shoved into the purgatory for all transitioning service members and veterans: human resources (HR). This is how it works: you don't get the other person excited, but they still feel obligated to help. After all, you are (or are about to be) a veteran. So, they say something nice to you via email and then loop in someone from HR. This person tells you, very politely, to apply for a job that's already listed online. They might even put the link to that job in the email response. How nice of them, right? Wrong.

If you get pushed off to a person in HR, then you did something wrong. Thank you for your service, and game over! You lost. Now you are competing with hundreds or thousands of other candidates. And you might be an amazing NCO, but you won't beat out all the other people from traditional industries with polished résumés and years of relevant work experience—not a chance. You will never hear back, and you will never know why.

	Education	Years of Experience	Last Job
Exceptional NCO Applicant	✗	0	n/a
Traditional Applicant	✓	5	Similar position

Going to HR purgatory is your punishment for failing to excite the other person. That's why it is such a key skill for anyone who wants to be a Startup NCO. The opportunities you want to pursue are all connected to specific people at specific companies. Those people will have to overlook a lot of reasons *not* to hire you. Excitement is the most powerful tool you have in the fight against the "Thank you for your service" response.

GETTING TO THE SECOND CALL

For now, we're focusing on how an introductory conversation leads to that excitement. There's a simple goal: the other person leaves the call thinking, *Wow, they are really sharp. They need to join our team!* The way to test whether you've succeeded is by asking them for an introduction to another person at the company. That is why the email template included a spot for you to put in that person's information. You have to make it simple and easy to get the next introduction. That starts with your follow-up email to that person once you are introduced. It's a great chance to be clear, concise, and focused. Here's an example:

> Hi [*First Name*],
>
> I am really excited to speak with you about your team at [*Company*]. Are you available for 30 minutes to chat this [*day*] at [*time*]? I won't waste your time. My goal is to learn enough about your goals and challenges to see if I can help.
>
> I know I'm not a traditional candidate for your team, but I've been working in these kinds of fast-paced environments for years. I work hard, make everyone around me better, and never stop learning.

Sincerely,
[*Your First Name*]
[*Your Phone Number*]

Most people in the military aren't comfortable sending emails like this. They don't like talking about their own value. As discussed in Chapter 7, they think of it as bragging or being salesy, even if every single word is true. You need to overcome that resistance and learn this skill as soon as possible. No one else is going to explain your value; that is your responsibility alone. And everything you do will need to match your claims if you hope to become a Startup NCO. For example, once the person confirms with a time, try this reply:

Great, thanks so much for confirming [*time*] on [*day*]!

In the meantime, please let me know if there's anything else you want me to read to prepare. So far, I've looked at the company website, read through some news articles, and skimmed LinkedIn profiles of the leadership team.

It seems like such a cool company. I can't wait to learn more!

Sincerely,
[*Your First Name*]
[*Your Phone Number*]

This is the kind of preparation that shows hunger, drive, and the ability to communicate. It helps the person feel excited about getting to speak with you. Most people, even the talented and educated ones, don't do this kind of prep work. Let the military brand work for you. Embrace the archetype of the hard-charging veteran who will bust through walls to accomplish the mission. Leaders want those kinds of workers on their teams. There is always a need for relentless, humble people who know how to get stuff done.

What happens after that second call? You follow up again, using the same technique to get introduced to yet another person at the company. Each time, you should be speaking with someone other than HR. Meanwhile, each person who talks to you then talks to their coworkers about you. The buzz starts to build. After that, your main job is to live up to the expectations of each new person you meet at the company. Let them think about you in terms of the positive military stereotype, but combined with a surprising degree of soft skills such as kindness, empathy, and a sense of humor. They will start to see you as someone who will inevitably be successful out of uniform. We'll talk a lot more about that in the next chapter on Deciding.

Your follow-up emails should look something like this:

Hi [*First Name*],

Thanks again for speaking with me yesterday. I love the idea of joining the team as a [*Job Title*] like you mentioned. The timeline works well for me, and the opportunity is similar to other roles I'm considering.

I'd enjoy meeting other members of the team to see if I'm a fit. You mentioned [*a potential contact's name*] and [*a different potential contact's name*]. Assuming the easiest way to connect is by email, would you mind introducing us? I can set it up from there and then circle back.

Sincerely,
[*Your First Name*]
[*Your Phone Number*]

Reread this email and think carefully about what makes it useful. It's meant to be read by others—specifically, the other people who work with the recipient. Why is that important? Because this email can simply be forwarded, reducing the burden on the other person to almost nothing. A Startup NCO knows to keep the conversation informal. Ideally, you will talk to everyone on a team before you get an offer for the job you want.

You need to use all your creativity to keep people engaged and excited about having you join the team.

That is why you write emails like the one above, not formal messages meant for someone in human resources. A more formal email will get you shoved into a backlog of tens of thousands of other people, and none of them will get the job. It is yours if you can stay relentless and disciplined.

BECOMING A GREAT DEAL

Cultivating conversations are not magic. They take advantage of societal stereotypes combined with basic human psychology. Imagine if you had the chance to buy something for fifty dollars that you knew was worth one hundred dollars. Wouldn't you do it? And wouldn't you feel great about the deal? Of course you would! This is the same dynamic you are trying to create in the minds of the people you meet. They should think of you as a great deal. Your skills are an ideal fit for the company. You are exactly what they need right now. And they lucked out by stumbling across you!

Once again, this strategy might sound too good to be true, but believe me: it works. It's a great example of that "running a marathon" type of hard work that will get you where you want to go. Follow the steps of the Cultivating phase, and within one month you will be having these kinds of conversations.

If Exploring is all about quantity, Cultivating is all about quality. A Startup NCO knows how to invest time wisely. A few minutes of preparation is fine for all those hundreds of Exploring conversations, but now we shift toward in-depth research. Every Cultivating conversation should leave the other person with the impression that you were surprisingly prepared and knowledgeable. That is part of what generates the excited energy that you need for your follow-up. A good rule of thumb is 1:1. Prepare for every minute of conversation with one minute of preparation ahead of time. A half-hour call would therefore require half an hour of time split between preparation (fifteen minutes) and follow-up (fifteen minutes).

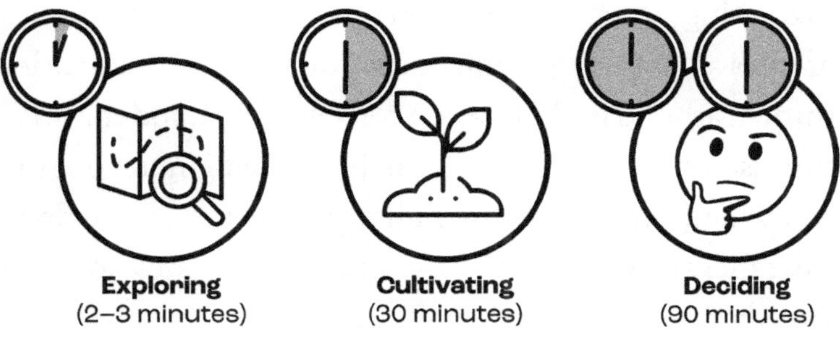

The Cultivating phase is when you start to increase your level of discipline and intention. You will have conversations with people who want to steer you toward opportunities at larger companies. Certain key words and phrases will keep popping up, often in reference to a job that you never even knew existed. This is when you follow the technique I mentioned during my own crucible: google every new word you hear, read about it, and then talk to people about it until it makes enough sense that you can find a way to connect your military experience to the key skills of that particular job.

GET IN WHERE YOU FIT IN

This part of the process gets increasingly complex. Let's fully unpack an idea that I touched on earlier: the larger the company, the greater the diversity of roles. Maybe you're being steered toward customer success, digital marketing, or channel partnerships. These are examples of roles at later-stage companies. You will hear these titles thrown around in Cultivating conversations, and they will confuse you for a while. Eventually, you will figure out the equivalent function in the military, allowing you to draw a parallel and even tell a story that demonstrates how you already have some of the skills needed for that role.

Startup NCOs do not pursue jobs at big tech companies, at least not if they are trying to make a name for themselves in the startup world. Resist the temptation to get steered toward these cool-sounding jobs at big companies that you've already heard of. If you want to join a larger, more stable company, then follow up on that conversation. But if you want to be a Startup NCO, do not let yourself get distracted by those opportunities.

This is an important point that we've already covered. There are many, many reasons why a Startup NCO avoids big tech companies. The hiring process is rigid and controlled by HR. The upward mobility is limited at larger companies, especially if you lack educational credentials. The teams move slowly, focus on process, and tend to be very politically correct. Merit and impact take a back seat to politicking, schmoozing, and other theatrical behaviors.

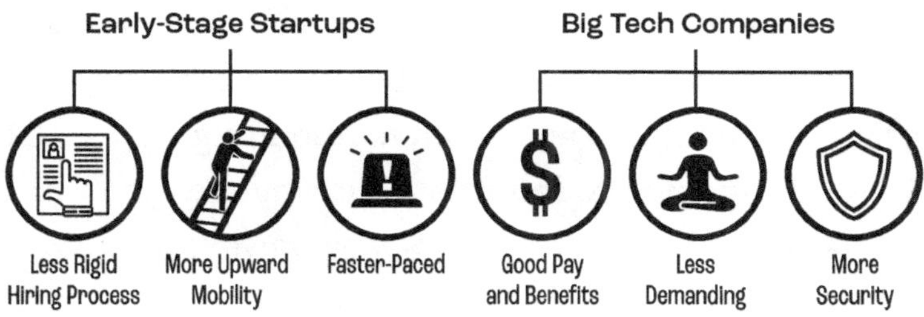

You may want to work at a big tech company someday. That is understandable. There are a lot of other people who do too. Some days, even I want to throw in the towel and get a job at a Meta or an Apple! There are obvious reasons to compete for those jobs: They pay well and have good benefits. You don't have to work anywhere near as hard. The companies don't run the risk of shutting down overnight. The work is more structured, more predictable, and less mentally demanding. If that is what you want, then go after it, and don't feel the need to apologize to anyone about it. Just remember how hard it is to compete with the hordes of professional knowledge workers who went to the "right" schools and have the "right" experience. I do not like our odds.

VETS HELPING VETS

There is one more issue to raise while we're thinking about how to successfully proceed through the Cultivating phase: other veterans. Veterans who have already joined a company will be informally connected or may have formed an employee resource group (ERG). These groups tend to focus on increasing the number of veterans who work at the company, raising awareness of veterans' issues, and so on. These objectives are fine, of course. But in practice, this kind of organizing tends to create a funneling effect. In companies with veteran resource groups, veteran applicants are usually steered toward specific jobs that aren't a part of the core business but have "veteran-friendly" managers and other support mechanisms.

It's hard to overstate the difference between an early-stage startup and a big tech company. That's why we're spending so much time on it here. For a true Startup NCO, a job at a larger company is what you take as a break between startups. It's a place to catch your breath, not to make your career.

MORE CONVERSATIONS = MORE OFFERS

We began this chapter with a two-part challenge:

1. Figure out whether you are interested in joining a specific startup. You do this by paying attention to the conversations that excite you.

2. If you do get excited about a particular opportunity, help the other person answer the question "Do I want this person to join my team?" with an enthusiastic "Yes!"

These actions create the momentum that will carry you through the Cultivating conversations in which you begin to better understand the work and the team. The more of these conversations you have, the more job offers you will receive. You may start with only one or two, but more will come. Job offers are an inevitable outcome of good Cultivating conversations. Now, we turn to the critical next step: deciding which offer to take.

TEN

DECIDING

The hardest decision for anyone leaving the military is who they will become. Who are you when you take off the uniform? For future Startup NCOs, this is an especially difficult question to answer. Many attractive options pop up during the Startup Crucible. The conversations in the Exploring and Cultivating phases open up new possibilities. Each path points toward a distinct future, however vague. After months of talking to different people, you'll find the world seems to be bursting with options. The sheer range quickly becomes paralyzing.

Contrast that feeling of energizing possibilities with the opposite experience of many people as they leave the military. Some transitioning service members choose continuity and comfort over novelty and challenge. They do not want to seek out new professional adventures. Their life after the military is a series of predictable chapters in the same old book. This is the potential tragedy of life as a veteran: if we allow our identities to stay rooted in our military service, then everything we do after leaving the military will be a footnote. We'll already have reached the climax, and now we'll be trying to drive a car by staring into the rearview mirror.

If you want to be a Startup NCO, then you are choosing to take a completely different approach. Of course you will be proud of your service, but it will not define you. You will retain and adapt the best of the military to launch into a fulfilling career, but as a Startup NCO, you fully expect to make an even bigger impact going forward. Your best days are ahead, not behind you.

But transitioning from the military is a lengthy, intensive journey. It takes years, if not decades. Anyone who tells you otherwise is deluding themselves or has not gone through it. Transition is a tough process of looking inward and outward, of knowing yourself and learning about others. The farther you get in your transition, the higher the stakes. That is why I didn't tackle this subject until now.

You can handle these sensitive subjects now that you have a better handle on the work that needs to be done.

DREAMS OF THE FUTURE

Your identity undergoes the most profound shift between the phases of Cultivating and Deciding. The stakes rise from almost zero to very high. It is easy to have a conversation with someone, but the closer you get to job offers, the scarier the situation becomes. Each job offer represents a totally different life.

Picture the future that goes along with a specific job. Try to imagine yourself living out each day. Reflect on the conversations you've had in order to include the details that make it seem real. Picture your boss, your new team members, your new home, your new office. Picture your new daily routine, from waking up and getting ready to coming home and decompressing. Pick a specific day in a specific month in a specific year. Maybe the weather is a lot different. If these things have not happened in your life already, imagine a scenario where you are married, have kids, or both.

For a Startup NCO, this is the moment when things start to click. The vague opportunities you've been discussing are now solidifying into concrete opportunities for a new life and a new identity. If they're the right fit, these futures should be desirable, interesting, and exciting to you. You should feel an immediate emotional tug toward these new versions of yourself. That is the most obvious test to assess whether a job is worth pursuing. Remember the question above: Who are you when you take off the uniform? This job should help you realize that vision of yourself.

It might seem like I'm overemphasizing your emotions. Believe me: I'm not. To become a Startup NCO, you must be on fire for your work. The sacrifices of the chaotic startup lifestyle must be worth it. That is the only way you will be able to do the thankless work that is necessary to excel. Otherwise, you'll half-ass the process, and then you won't stand out enough to get or keep a great job at a great startup.

And it is not just about your career. You also need to be on fire for your entire life. This is a much wider lens to look at your future. Does this job let you live a better life on the terms that matter to you? Are you building a product that you believe in? Are you far from family when you want to be close, or the other way around? Are you sacrificing higher compensation at a time when you need to save? Are you signing up for a lot of travel with young children at home?

Remember, as a Startup NCO, you are a nontraditional candidate. You're competing with people who have the "right" skills and experiences for the job. You are fully capable of beating out others and earning the position, but only if you push yourself to a higher level. You need to draw energy from that vision of your future life. That energy will power you to go the extra mile when it makes all the difference. It will make you the obvious choice for the people who are offering you a job at their startup.

If you do not go the extra mile, you will not get that great job. All of us after 9/11 have heard "Thank you for your service" too many times. Without meaning to internalize the well-meaning gratitude of civilians, we have. We think we are just a little better than we really are. We underestimate the work required to adapt ourselves to a new role, a new industry, and a new life. In other words, a self-sabotaging part of us whispers that we are entitled to the job. That voice tells us that we do not need to go above and beyond.

ALL JOBS ARE NOT CREATED EQUAL

Tragically, you *will* get a job with that attitude. The problem is the job is likely to be a bad one. There is no shortage of bad jobs—believe me. And that might be fine if you want to get shoehorned into a role with little room for upward mobility. But Startup NCOs do not compete for those types of jobs, because they in no way offer the type of work or life we want. Instead, we identify, pursue, and get offers for amazing jobs that are in high demand. We don't want to miss out on those opportunities because we took the first offer.

The Startup Crucible helps you avoid these kinds of mistakes. In the Deciding phase, things get real. You are following up with people to get them to refer you to people who are hiring for specific open positions. Your goal is to get at least five phenomenal job offers within a few weeks of each other so that you can select the one that is most appealing to you. That's why it is so important to imagine your future self in each specific job, working with each specific team, doing specific work. That exercise will help clarify which opportunity is right for you.

There is no obvious dividing line that indicates you've finished the Cultivating phase and are now onto Deciding. The transition happens naturally as you meet more people, learn to listen for what matters to you, and get better at telling your story. You can't push yourself into Deciding; it's something you are pulled into by others.

None of this means you should pursue every opportunity in front of you just because you think it will turn into a job offer. Startup NCOs make the most of their available resources, which means being strategic. We cannot be everywhere at once. In the Deciding phase, we need hours of preparatory research for each job opportunity. That's impossible if we go after everything. There are only so many hours in the day.

At this point, it is time to apply some filters to narrow down the breadth of what you are considering. A few weeks ago, you were worried about having enough people to talk to, but now you need to become more selective. That is testament to the effectiveness of the Startup Crucible. There are specific dimensions to consider, and they all link back to common aspects of your identity after leaving the service: family, boss, team, location, pay, and, last of all, work.

Family

What do you owe the people you love? Sadly, family considerations often come last when people think about jobs. The opposite approach is almost always the correct one. People in your family have made many sacrifices so that you could put military service above the other considerations in your life. Now might be the time to prioritize others over yourself. Slow down and take the time to discuss each opportunity with your spouse, parents, siblings, children, and so on.

You can't read anyone's mind, so without having these conversations, there is no chance you will know exactly what others want. They live in their own heads and, like you, may sometimes have a hard time explaining what they want or why they want it. There will be breakthrough moments when you listen carefully to them. Plan out several blocks of time over a series of weeks or even months. Be intentional about how you approach these conversations. Clarify how much you care about them and how important it is for you to hear about their expectations for this next phase of your life together. Talking about important decisions with the people who love you is a habit that will serve you for the rest of your life.

Boss

Who will you work for? This decision is the most important one you will consider for any job. The fate of a startup directly correlates to the quality of its leaders. The government is a stable environment, the exact opposite of a startup, which sets the wrong expectations for veterans. If you join a startup with fifty employees, for example, a few years later it will either employ hundreds to thousands of people or zero. The leaders of that startup will be the major factors affecting the outcome. This stark win/lose dynamic is the nature of startups, so we have to choose wisely.

This decision is critical but tricky, because you will probably suck at it. Choosing a boss is a skill, and we don't get to make that choice in the military. The people leading us are effectively set in stone. We therefore lack the basic competency that private-sector folks enjoy by the time they're in their second or third job. Without the skill of boss-picking, we will naturally focus on the aspects that are easier to measure, such as pay. But there is a reason that pay is not even near the top of our list: the right boss can make all the difference.

Team

Who will you work with? This is another decision that you've probably never had to make. Choosing your teammates is critical, though, because of the huge impact they will make on your quality of life on a daily basis and, over the longer term, your career trajectory. Many NCOs are used to being one of a few high performers, forced to make up for others who are phoning it in. As a Startup NCO, you want to find a team where you are impressed—if not intimidated—by everyone around you. Those people will push you to be better, and you should thank them for it.

Another advantage of working in startups is the opportunities that will come your way. A Startup NCO constantly gets offers to join companies that are being started by former coworkers. This is the natural state of the startup world. People see business-shaped opportunities and build teams to get after them. The better the team, the more likely it is that folks will leave on a regular basis. Every person who leaves will reach back to recruit the best talent. That should be you, giving you yet another opportunity to consider.

Location

Where will you work? Most people either stay where they currently live, move to be near a college or university, or relocate to be near family. These are understandable default settings. As Startup NCOs, however, we need to take a closer look at the options. Imagine that you wanted to join Special Forces and tried to do all the training on your own, without investing in the right equipment or coaching. You would be much more likely to fail during selection if you ever qualified at all.

Deciding to pursue life as a Startup NCO is similar. You will need to learn to use the tools of the trade properly and find high-quality mentors who have worked in many startups at different stages of growth. These "player-coaches" are a critical part of any startup community. They can draw from decades of tacit knowledge to advise you, offering critical, timely insights that you will not figure out on your own. Startups are difficult in many ways, and it takes time and hard work to grapple with each challenge as it arises. Choose a location that will help you learn from others with real-world experience.

Pay

How much will you make? It is really easy to screw up conversations about how much you'll get paid. People leaving the military are used to reasonable pay with good benefits and extra allowances. That is not what you can expect at a startup. Instead, think critically about what you actually need to handle the basic necessities—and then how to participate in the upside if the company ends up being really successful. That means you will need equity as a part of your overall compensation.

You would mess up this conversation if you had to handle it on your own. Fortunately, you do not have to do that. This is one of the key reasons that choosing the right boss comes first. You should work for someone who is trustworthy and understands your value. That person—even though they work for the startup—should help you navigate the negotiation process. Of course you can discuss the situation with family, friends, and mentors. The main thing to keep in mind as a Startup NCO is that you should trust the person who is about to become your new boss, particularly around the issue of pay. If you don't, then you should never even consider taking the job.

Work

How will you work? It may seem strange to leave this until last. After all, that's what a job is, right? Well, yes and no. A Startup NCO understands that everything about a high-growth business changes—that is the very essence of growth. It stretches the team, opening up new opportunities all across the company. A high-performing member of a fast-growing team will change "jobs" at least once a year, if not more frequently. That's why you should clearly understand the work that's required and how it fits into the business, but not obsess about it. Show up, crush the job you have right now, and let the startup's growth take care of the rest.

FOLLOWING UP

Your future as a Startup NCO is tightening and brightening. As you narrow down the range of options, you invest more time preparing for every interaction. You must spend enough time considering each job to stand out in interviews, and that comes from asking insightful questions that clearly show how much effort you've put into preparation. Armed with this information, you can follow up with prospective bosses like this:

> Hi [*First Name*],
>
> Thanks again for the chance to speak with your team. I'm even more excited about the opportunity now, and

I'm convinced I am exactly the right person to fill this near-term role while growing along with the company. The main things that jumped out at me are:

- [*immediate opportunity to pursue*]
- [*key problem(s) that need solving*]
- [*noteworthy trend from conversation(s)*]
- [*attention-grabbing insight(s), especially quotes from your earlier conversations that stuck out to you*]

I'm deciding among several opportunities in the next week or so. [This Company] is at the top of my list. Can we hop on a call this [Day] at [Time] to discuss?

Sincerely,
[*Your First Name*]
[*Your Phone Number*]

You're now one or two conversations away from receiving a verbal job offer. That is your potential boss telling you that they are going to offer you a job. This conversation may also cover details such as the title of the role, the start date, the location, and the compensation package (which includes salary, benefits, and possibly company stock). This step is followed by a written offer, usually via email as a Word or PDF attachment. If the company is a true early-stage startup, that's really all it takes, because they won't have formal processes for hiring and onboarding people. For more mature startups, there will be a specific person who discusses the offer with you, usually someone from HR. That's the time to ask questions about pay, benefits, vesting of stock, relocation, start date, and anything else that's on your mind.

THE CHOICE IS YOURS

Remember your goal is to select the best job offer from at least five, all within a short time frame. Five options represent the optimal range. You don't want so many possibilities that you feel paralyzed—that tends to happen beyond five, which is why it is our goal for the Startup Crucible—but you also don't want to feel forced into accepting something that isn't a compelling opportunity.

As you get your offers, think through the major factors listed above, along with any other elements that are especially important to you. I recommend setting yourself a deadline between three and seven days in the future, although you can take longer if you expect more offers to come in soon. Tell each person who offers you a job what your deadline is and explain that you are deciding among multiple offers. Sometimes that simple fact will lead them to improve their offer with higher pay, a signing bonus, more flexible work arrangements, or more equity.

Once you decide, do not look back. Commit to the decision, even if it is a close call between two great opportunities. Then, make sure you notify everyone else about your decision *before* the deadline you set. Do not burn any bridges with people just because you are not working together. You should thank everyone who offered you a job and everyone else you talked to at each company. You never know when you'll need one of them to open a door for you in the future. Startups are risky, and reputation is everything.

Congratulations—and welcome to the startup life!

ELEVEN

ONBOARDING

The Startup Crucible doesn't end when you accept an amazing job offer. There is still one final phase, which brings together everything we've discussed: Onboarding. The first few months at your first startup offer a unique opportunity that will never happen again. You get to take everything you learned from your time in uniform, plus the Startup Crucible approach, and focus on doing as well as you can in the first ninety days of your new job. This is your chance to prove to everyone that you can, in fact, thrive—and it completes your transformation into a Startup NCO.

There are sky-high expectations for you to meet. The main reason you were offered this job was your incredible performance during the interview process. Everything you did, from the preparation to the follow-up, made a positive impression on every single person you met. You set yourself apart from the crowd. Your new boss decided to take a risk on you, instead of going with a more traditional candidate. They will now be asking themselves if they made the right call. In the first ninety days, your job is to ensure the answer is a resounding "Yes!"

Being a Startup NCO does not mean trying to mimic everyone else on the team. You bring a new perspective and new skills to the table. No one wants you to be the same as everyone else. Your new colleagues want you to fit in, of course, but also to raise the bar. You were hired for a reason. They want to see you bring the same magic that you demonstrated during the interviews to transform the company from within. They also want to see you instill these same abilities in others.

To succeed, you must be laser focused during the Onboarding phase. It's tempting to let your enthusiasm get the better of you. Many people start new jobs with an exhaustive list of all the things they want to do and try to knock them out, one at a time. That is the wrong approach. Remember the inescapable reality of startups: they are trying to become high-growth businesses, which requires incredible amounts of work and constant adaptation. That chaotic reality threatens a lot of people, but not you. In fact, it plays to your strengths. Your instincts as a Startup NCO, which you have already begun to adapt and refine during your Startup Crucible, will guide you a lot of the way.

What you need is one simple strategy: avoid making really big mistakes. The rest will take care of itself. Trust that you can pick up the culture, the tools, the jargon, and everything else as you go. That type of learning is natural. You will be like a dry sponge, just as you were at boot camp. Manage your expectations: it will take months to get familiar with the various ways of working at your new company.

Here are the big mistakes that will hinder you as a new employee at a startup:

- Not trusting your instincts
- Getting overwhelmed with information
- Avoiding big projects
- Being siloed within a team
- Not asking enough questions
- Forgetting to do regular check-ins
- Not documenting the journey
- Not mentoring others

These are all important to think through, and you will need to address each of them to successfully onboard and build momentum in the job. As you review the following explanations, remember that the goal is to avoid these common mistakes. Think about the ones that you *specifically* are likely to make. Which of these mistakes will tempt you? Spend time identifying each mistake and then overwriting it with the actions you will take instead.

NOT TRUSTING YOUR INSTINCTS → TRUST YOUR INSTINCTS

You are joining a team that is already impressive. They have a variety of skills and experiences that allow them to be successful in many ways, but they are not you. You are a Startup NCO. Your perspective and skills are unique; that's a big reason why they hired you in the first place! As you start to get your arms around the culture and the work, you will see obvious improvements that need to be made. Do not censor yourself. You see things differently, which allows you to identify problems that others don't even notice. Talk to your boss whenever you have a concern triggered by your instincts. There is no harm in having the conversation.

GETTING OVERWHELMED WITH INFORMATION → TRIAGE INFORMATION

You're used to lots of military processes, policies, and procedures. The company will ask you to learn an entirely new set of them. That means new forms, new tools, new language, and so on. You will start to lose some confidence unless you expect this and prepare for it. It's critical to get comfortable with letting some things slide. Everything you're doing is not urgent and important. Focus energy on learning about the big stuff, rather than fixating on the little stuff. Startups do not have the support structures of big companies. It is up to you to ensure you have a clear understanding of your top priorities. When people throw more random information at you, ask if it's okay to circle back in a few months once you are more firmly placed in your job.

AVOIDING BIG PROJECTS → JOIN BIG PROJECTS

Based on the conversations during your Startup Crucible, especially the Cultivating and Deciding phases, you should have some pretty good ideas about the key things you need to do in your first ninety days. Do not lose sight of these efforts as you get started. Things move fast at startups. No one needs to ask permission to jump in and help. Talk to the person who hired you to find out who runs each big project (if you don't already know). You do not always need to ask to be assigned to these projects. In many startups, it's okay to just show up at the meetings, listen carefully, and follow up with individual people to figure out how you can help. Make sure to keep your boss informed if you decide to join a big project, but otherwise demonstrate the bias toward action that they are expecting.

BEING SILOED WITHIN A TEAM → MEET EVERYONE YOU CAN

If you truly embraced your Startup Crucible, you will be a mini celebrity at your new company. People will want to meet this military person who somehow got hired despite having none of the traditional qualifications. You are an outlier. Use this strong personal brand as an excuse to connect with as many people as you can, ideally one person per working day for at least the first month. Most people will agree to meet for lunch or coffee during a midmorning break. You can learn a lot from their perspectives on the company, the market, the product or service, and especially the team you're joining. Remember that every person you meet will eventually leave the company and go on to do great things somewhere else. In other words, they are not just current coworkers; they are potentially the source of your next great job.

NOT ASKING ENOUGH QUESTIONS → CLARIFY COMMANDER'S INTENT

You do not want to be the sort of person who works super hard on a project, only to produce something that totally misses the mark. This happens all the time, especially at startups. Many initiatives are overcome by events. Startup NCOs therefore need to be intensely strategic, testing hypotheses with minimal effort. Whenever kicking off something new or joining a team, make sure you understand the "commander's intent." Research the background and goals of the project, prepare a list of open-ended questions, and find a time when the person in charge isn't too busy. Move out once you get your answers, not before.

FORGETTING REGULAR CHECK-INS → GET REGULAR FEEDBACK

You will make a lot of mistakes early in your Onboarding. This is totally natural. Do not try to avoid making little mistakes. This is a risk-averse mindset that will undermine your ability to create value. Instead, focus on avoiding big mistakes, while also getting feedback from others. *How well was that information presented? Did I format this deliverable the right way?* Feedback is most effective when you receive it on a regular basis, ideally once a week with team members. This rhythm gives you time to do real work between check-ins but doesn't let you go for so long that you let mistakes become bad habits.

NOT DOCUMENTING YOUR JOURNEY → REFLECT DAILY

Take notes as you become a Startup NCO. You'll be surprised how much you change in a short period of time. Insights that seemed life-changing last month will be forgotten as you get after some big new project or master an amazing new tool. A few minutes at the end of each day jotting down notes to yourself is enough. Even a short period of reflection solidifies your learnings from that day while also giving you some space to think about tasks you need to accomplish tomorrow. If you do not want to do any written self-reflection, make sure to share your experiences by talking with family and friends.

NOT MENTORING OTHERS → INVEST IN OTHERS

There are two groups of people you need to start mentoring: coworkers and other NCOs leaving the military.

Each person at your new company is at a different stage of their career, with a unique blend of experience and interests. It is impossible to predict what your coworkers will need from you, but you will be glad you started these relationships by finding ways to help. You will have a lot to learn from your coworkers, of course, but do not let the relationship be a one-way street. You have plenty to offer them as well.

When it comes to other NCOs, you have a special responsibility to find the ones like you and help them start down the path to becoming a Startup NCO. You can probably rattle off the names of two or three people like you right now. Reach out to share your journey, ideally a little bit at a time. Occasional texts plus a monthly fifteen-minute call will be more valuable than a single two-hour conversation with no follow-up. Regular contact lets you get to know them better, and they will get to see how quickly you transform. Your journey will be hugely motivating.

Everything about your first startup will be uncomfortable at first. There is no way around that, unfortunately. You will have your own ways of dealing with that discomfort. Just make sure you are aware of your worst military stress behaviors so that you can notice and address them. Also note that many NCOs need to adjust their posture and volume in the private sector. Outside of the military context, their typical communication style can seem aggressive rather than confident. There is nothing wrong with smiling, saying "please" and "thank you" to coworkers, and otherwise acting as if life is pretty great. Because guess what? You're out of the military now, working at an amazing company with incredible people to make the world a better place. So yes, smile!

CONCLUSION

Joining the military is right up there with graduating from school, losing a loved one, and becoming a parent for the first time. It is a huge, transformative event that sends ripples across the rest of your life. Leaving the military gets a lot less attention but is probably more impactful for many of us. The path you choose when you take off the uniform will affect every other part of your life from here on out. This fact can be intimidating at times, but it is better to acknowledge reality than delude yourself and stumble in civilian life.

Being an NCO in the United States military is a big deal. You are in the top 25 percent of all Americans in terms of work ethic, adaptability, and grit—merely by reaching that level of responsibility. Assuming you did well as an NCO, you easily place in the top 5 percent of the entire population. That means you stand out as the top candidate in a group of twenty other people. And that is before you factor in the path you choose for yourself after taking off the uniform. The greater the goals you set for yourself, the more of your potential you will reach. Your ambition, properly channeled, will help you go far.

Success is far from guaranteed, though. There are countless examples of amazing soldiers, sailors, airmen, and Marines who floundered after they left the military. Despite their world-class performance within a more structured environment, many do not transition successfully. Their lives suffer as financial prospects dim, relationships strain, and their self-belief falters. They start looking backward toward an imaginary past and stop working to build a better future for themselves and their families.

Many of us worry, rightly so, that we will end up a cautionary tale. That's why we are tempted to choose a different, safer path. And there is nothing wrong with choosing a career in local law enforcement, or the federal government, or a big corporation—not if that is truly what you want. But if you sense that you are destined for something different and more challenging, then you should pursue that larger destiny.

You chose to take a harder road when you joined the military. You did not quite know what you were getting into, but you answered the call. Maybe you have that same feeling now, that same sense of a better life that is worth working for. Deciding to become a Startup NCO is one of the greatest goals you can set for yourself. You are choosing to work hard to build your life into something even better after the military. You are not just writing a new chapter in a book about your military service—you are writing an entirely new book about yourself!

Success does not happen by accident, even though you already possess the skills you need. It takes a strategy, a lot of hard work, resilience in the face of setbacks, and the courage to let go of parts of your identity. You must embrace the Startup Crucible, molding yourself into someone who fits well into a fast-growing, dynamic startup.

As you finish this book, there is a first step for you to take now. If you do this one thing, you will dramatically increase your odds of success. Create a daily calendar event or a reminder on your phone. Set it for the first thing in the morning or immediately after work, depending on your schedule.

Set aside an entire hour, and label this block of time "Startup Crucible." This is your time to invest in yourself and your future. Use it to get set up on LinkedIn and start reaching out to connections as described in the Exploring chapter. Use it to have follow-up conversations and track your learnings in a Google Doc as described in the Cultivating chapter. Use it to get job offers as described in the Deciding chapter. And use it to crush your first ninety days as described in the Onboarding chapter. Do not skip a single day. Embrace the discipline.

Finally, some people will be happy that you are taking on a difficult challenge, but not everyone. You will have to face family members, friends, and even your spouse or children who might want you to select a safer path. There are plenty of difficult conversations ahead. Prepare yourself by clarifying your vision for yourself and your family. Help them see themselves in that brighter future.

Fortunately, you will not have to go through this journey alone. If no one else supports you, I will. It took a lot of mentors for me to grow into a Startup NCO. I am honored to be one of yours. Reach out to me on LinkedIn to start the conversation. If you search for my name, I will pop up immediately. Let me know what sort of help you need, and I promise I will get back to you.

Good luck with your Startup Crucible, and God bless you.

RESOURCES

Note: If you're reading a print book and want clickable links for all of these resources, head over to *http://williamtreseder.com/startupnco*, or scan the QR code below.

READING LIST

Zero to One: Notes on Startups, or How to Build the Future, by Peter Thiel

Shoe Dog: A Memoir by the Creator of Nike, by Phil Knight

Build: An Unorthodox Guide to Making Things Worth Making, by Tony Faddell

Inspired: How to Create Tech Products Customers Love, by Marty Cagan

The E-Myth Revisited: Why Most Small Businesses Don't Work and What to Do About It, by Michael Gerber

"Navigating the Hiring Process" by Misha Chellam, with insights from Katie Hughes

LINKEDIN

Premium Business subscription for veterans: *https://bit.ly/LIforVeterans*

Premium subscription for active-duty personnel: *https://bit.ly/LIforAPs*

http://williamtreseder.com/startupnco

ACKNOWLEDGMENTS

It takes a village to raise an author. There are many people who deserve my gratitude and praise for their contributions to this book. My mother, God rest her soul, made me feel I had something worth saying. My father, an NCO himself, made me earn the privilege of saying it.

My enlisted buddies who are Startup NCOs—folks like Brendan, Dan, Paul, Justin, Dillon, Lydia, and Ross—walked the path alongside me. Our parallel experiences made me think there was a path for others. Our conversations convinced me this topic was worthy of a book.

My editor, Chas, accompanied me on this yearslong journey with grace and wit. He was always ready to talk through a particular issue or ensure I found a better way of capturing an important insight. This book would not have happened without him.

My children, Zoë Asta and Johan, keep me on my toes. They were just learning to write stories as I finished up this book. I hope I inspire them someday the way they inspire me today.

My wife, Dara, is my person. There is no other way to say it. She matches my energy in every way. I cannot imagine life without her, and that includes all the experiences that are captured in this book. I love you, My Queen.

Finally, I want to express my heartfelt gratitude to the NCOs of the past, present, and future. You all are a crazy tough bunch. You know when to suck it up and when to lighten the mood. You are the beating heart of our military. It is an honor to be among your ranks.

ABOUT THE AUTHOR

William Treseder is cofounder and chief operating officer at BMNT, a government innovation consultancy. In this role, William drives operational excellence for public-sector customers ranging from the National Security Innovation Network and Defense Logistics Agency to the Advanced Research Projects Agency for Health.

William is also the chief strategy officer for the Marine Innovation Unit. In this capacity, he works on the organizational structure and mission set with a focus on transformation. He is also (probably) the oldest sergeant in the Marine Corps, date of rank: 1 May 2005.

William previously served in product development and entrepreneurial education roles. His career includes formative projects with key defense innovation capabilities, such as Hacking for Defense, Kessel Run, NavalX, and the Defense Innovation Unit.

William has written extensively on public-sector innovation. His publication credits include *TIME*, *Harvard Business Review*, *Boston Review*, and *Breaking Defense*. He's also the best-selling author of *Reset: Building Purpose in the Age of Digital Distraction*. He holds a degree from Stanford University and an executive certification from Stanford's Graduate School of Business. He lives in Silicon Valley with his wife and two children.

www.ingramcontent.com/pod-product-compliance
Lightning Source LLC
Chambersburg PA
CBHW030505210326
41597CB00013B/807

* 9 7 9 8 9 9 1 2 0 3 9 4 4 *